T0155999

SpringerBriefs in History of Science and Technology

More information about this series at http://www.springer.com/series/10085

John M. Steele

Rising Time Schemes
in Babylonian Astronomy

 Springer

John M. Steele
Department of Egyptology and Assyriology
Brown University
Providence, RI
USA

ISSN 2211-4564   ISSN 2211-4572   (electronic)
SpringerBriefs in History of Science and Technology
ISBN 978-3-319-55220-0   ISBN 978-3-319-55221-7   (eBook)
DOI 10.1007/978-3-319-55221-7

Library of Congress Control Number: 2017933556

© The Author(s) 2017
This work is subject to copyright. All rights are reserved by the Publisher, whether the whole or part of the material is concerned, specifically the rights of translation, reprinting, reuse of illustrations, recitation, broadcasting, reproduction on microfilms or in any other physical way, and transmission or information storage and retrieval, electronic adaptation, computer software, or by similar or dissimilar methodology now known or hereafter developed.
The use of general descriptive names, registered names, trademarks, service marks, etc. in this publication does not imply, even in the absence of a specific statement, that such names are exempt from the relevant protective laws and regulations and therefore free for general use.
The publisher, the authors and the editors are safe to assume that the advice and information in this book are believed to be true and accurate at the date of publication. Neither the publisher nor the authors or the editors give a warranty, express or implied, with respect to the material contained herein or for any errors or omissions that may have been made. The publisher remains neutral with regard to jurisdictional claims in published maps and institutional affiliations.

Printed on acid-free paper

This Springer imprint is published by Springer Nature
The registered company is Springer International Publishing AG
The registered company address is: Gewerbestrasse 11, 6330 Cham, Switzerland

Acknowledgements

I thank the Trustees of the British Museum for permission to study and publish tablets in their collection, Mathieu Ossendrijver for sharing photographs of A 3414 and A 3427, and Christine Proust for information on the tablets in Istanbul. My work on the rising time tablets was made possible by a Franklin research grant from the American Philosophical Society and the final manuscript was prepared during my time as a fellow of the Institute of Advanced Study at Durham University.

Contents

Chapter 1
Introduction

Abstract Babylonian rising time schemes relate positions at or relative to stars which culminate either as the sun rises or sets on particular days of the year or as given points on the ecliptic rise across the eastern horizon. These schemes were first identified in cuneiform texts in the mid-20th century and were initially assumed to be related to the functions for the length of daylight in the mathematical astronomy of the Hellenistic period (the so-called 'ACT astronomy'). Subsequently, their relationship with earlier types of Babylonian astronomy has been recognized. This chapter outlines previous research on the rising time schemes and provides an overview of the sources to be studied in this work.

Keywords Babylonian astronomy · Ecliptic · Equator · Oblique ascension · Rising time · Zodiac

1.1 General Introduction

Developing a method to model the effect of the obliquity of the ecliptic to the celestial equator was an important problem in ancient astronomy. The obliquity is a consequence of the tilt of the Earth's axis by roughly 23° away from perpendicular to the plane of the Earth's path around the Sun. As a result, at a given geographical latitude the length of arc of the equator which corresponds to a given stretch of the ecliptic varies depending upon where that stretch is in the ecliptic (see Fig. 1.1). For example, the arc of the equator which crosses the meridian during the period when the zodiacal sign of Aries rises over the eastern horizon is considerably smaller than the arc of the equator corresponding to that of Libra as it rises. Distances on the celestial equator correspond to intervals of time: one day corresponds to the 360° circle of the equator, and therefore 1° corresponds to 4 min of time. Thus, the length of time it takes for a sign of the zodiac to rise over the eastern horizon also varies between signs. The obliquity thus causes the length of daylight, which corresponds to the rising of a 180° stretch of the ecliptic, and the height of the Sun at midday, to vary across the year, giving rise to the seasons: in summer, the Sun

© The Author(s) 2017
J.M. Steele, *Rising Time Schemes in Babylonian Astronomy*, SpringerBriefs
in History of Science and Technology, DOI 10.1007/978-3-319-55221-7_1

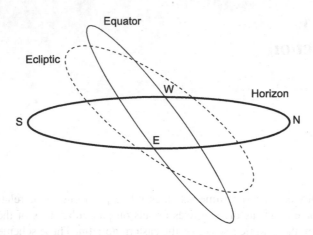

Fig. 1.1 The ecliptic and the celestial equator as viewed from Babylon in the winter. The thick line marks the horizon as seen from Babylon. The celestial equator forms a great circle which crosses the horizon at the east and west points. The complete circle of the equator crosses the east point every 24 h. The dotted line marks the ecliptic. The ecliptic is a great circle along which the sun appears to move, completing a full circuit in the course of the year. The ecliptic may be divided into twelve 30° sections known as zodiacal signs

will be above the horizon for longer and reaches a higher noon-altitude than in the winter.

Determining the relationship between arcs on the ecliptic and the corresponding arcs on celestial equator, a problem often referred to as determining the "rising times" of zodiacal signs (or parts thereof) or "oblique ascensions", is a basic problem of spherical trigonometry requiring knowledge of the geographical latitude, the angle of intersection of the ecliptic and equator, and the position on the ecliptic of the spring or autumn equinoctial point where the ecliptic and equator cross. This problem was solved theoretically by Greek astronomers and a table of rising times is found, for example, in book II, Chap. 8 of Ptolemy's *Almagest* (Neugebauer 1975: 34–35). A non-trigonometrical approximate solution to this problem, almost certainly based upon knowledge of Babylonian arithmetical astronomy, is given by Hypsicles in his *Anaphoricus* (Montelle 2016).

This study investigates various Babylonian texts which are concerned with the problem of what we would call the oblique ascension. These texts relate either the rising of signs or parts of signs of the zodiac with intervals between the culmination (i.e., crossing the meridian at their highest point) of a select group of stars known as "*Ziqpu* Stars", or give statements of positions relative to *Ziqpu* Stars which culminate at sunrise and/or sunset on specific dates in the year. As I will demonstrate, all of these texts attest to a single basic scheme. Furthermore, there is good reason to conclude that the date-based version of the scheme came first and was subsequently reworked into the zodiac-based version. Thus, although I continue to refer to this scheme as a "rising time scheme", in accordance with earlier literature and for lack of an obviously better term, it must be remembered that what is rising or setting may be either a sign of the zodiac or the sun on a given day.

Two more remarks are briefly worth making here: First, the rising time scheme incorporates several ideas and conventions which are most clearly formulated in the early Babylonian astronomical work MUL.APIN[1]: the 2:1 ratio for the length of daylight at the solstices,[2] the use of the schematic 360-day calendar, and the placement of the solstices and equinoxes on the 15th day of months I, IV, VII and X. The use of these conventions places the rising-time scheme within the genre of what I have termed "schematic astronomy" in Babylonia.[3] Secondly, all of the texts which include the rising time material are descriptive rather than procedural. Babylonian theoretical astronomy is often characterized as being procedural: texts give instructions for how to calculate astronomical phenomena rather than explicitly presenting a theoretical model for those astronomical phenomena. However, the rising time texts, in common with most other texts of schematic astronomy, present astronomical data without any instruction for how that data is to be used. I will elaborate further on these two characteristics of the rising time texts below in Chap. 6.

1.2 Previous Work

Babylonian use of the concept of rising times was first recognized by Neugebauer in his analysis of the schemes for the length of day in the System A and System B lunar theories. In his 1936 paper "Jahreszeiten und Tageslangen in der babylonischen Astronomie", Neugebauer showed that underlying column C of the lunar systems were arithmetical schemes with increasing and decreasing sequences of either constant difference (in System A) or with constant difference except for the middle difference which was twice as large as the others (System B) (Neugebauer 1936: 530–538). Neugebauer noted that these schemes were identical with schemes for the rising times known from Greek and later astronomy, but allowed for the possibility that they might simply have been convenient arithmetical schemes used by the Babylonians rather than necessarily have been understood as rising times. In 1953, Neugebauer published another paper, "The Rising Times in Babylonian Astronomy", in which he was able to demonstrate from the procedure texts ACT

---

[1]MUL.APIN is edited and translated by Hunger and Pingree (1989), which also includes a brief commentary. A new edition and commentary is being prepared by H. Hunger and the present author.

[2]The appearance of the 2:1 ratio in the "microzodiac" rising time texts was first recognized by Rochberg (2004).

[3]On this term, see Steele (2013) and the discussion in Sect. 2.2 below.

200 and 201 that the System A scheme at least did indeed represent rising times (Neugebauer 1953).

The rising times as they appear in System A and System B are presented in Table 1.1. In each case, the values are taken to begin at 10° within the sign (System A) or 8° within the sign (System B) and to extend to 10° or 8° within the next sign. The rising times are presented without units and, unlike all other instances where rising times appear in Babylonian astronomy, are not stated by reference to the culmination of the *ziqpu*-stars.

Shortly after Neugebauer's discovery, Schaumberger identified two groups of texts which relate to the rising times: (i) texts which divide each sign of the zodiac into twelve "microzodiac" parts and gave positions at or behind *Ziqpu* Stars which culminate when the microzodiac rose across the eastern horizon, and (ii) a text which he interpreted as a rising and setting calendar ("Aufgangskalender") (Schaumberger 1955). In the first group, Schaumberger studied three texts: A 3427, LBAT 1499 and the badly preserved U 196.[4] He concluded that the rising times presented in the first two texts were in agreement with the scheme Neugebauer had identified from System A but that because of the damage it is not possible to be certain whether U 196 agreed with System A or System B. In the second group, Schaumberger correctly identified U 181a, b, c, and d, and A 3414 as fragments from the same tablet. The tablet gives positions at or behind *ziqpu*-stars for setting (ŠÚ) and rising (KUR) on each day of the year. As I will show in Sect. 3.3.1, Schaumberger's incorrectly placed the five disconnected fragments in his reconstruction and as a consequence misunderstood the underlying scheme.

Significant progress was made in understanding the texts which give the rising time scheme using the microzodiac by Rochberg in her 2004 article "A Babylonian Rising Times Scheme" (Rochberg 2004). Rochberg reedited two of the texts studied by Schaumberger, A 3427 and LBAT 1499, and added a third, LBAT 1503, containing the same type of material.[5] Rochberg convincingly demonstrated that the three texts did not follow either the System A or the System B scheme but rather a much simpler scheme in which the rising time equals 20° for six signs of the zodiac and 40° for the other six signs. Her reconstructed scheme is presented alongside the System A and System B schemes in Table 1.1. Rochberg further demonstrated that this scheme would produce a zigzag function for the length of daylight with extrema in the ratio 2:1, similar to the daylength schemes in MUL.APIN.

---

[4]Schaumberger almost certainly used a copy of photograph of U 196 sent by F. R. Kraus to Neugebauer in 1945; in addition to the poor preservation of U 196, the photograph is badly lit making the tablet even harder to read.

[5]Not being able to collate U 196 and acknowledging the uncertainties in Schaumberger's transcription, Rochberg sensibly decided to omit this tablet from her study. In addition, Rochberg omitted the small fragment BM 77242, published by Horowitz (1994); this tablet can only be understood once the whole rising time scheme has been reconstructed.

Table 1.1 Summary of the rising time schemes identified by Neugebauer in lunar Systems A and B and by Rochberg in the microzodiac texts

Zodiacal sign	System A (°)	System B (°)	Microzodiac scheme (°)
Aries	20	21	20
Taurus	24	24	20
Gemini	28	27	20
Cancer	32	33	40
Leo	36	36	40
Virgo	40	39	40
Libra	40	39	40
Scorpio	36	36	40
Sagittarius	32	33	40
Capricorn	28	27	20
Aquarius	24	24	20
Pisces	20	21	20

1.3 Types of Texts Containing Material Relating to the Rising Times

Table 1.2 summarizes the sources studied in this work. I do not consider the evidence for rising times in lunar System A and B, which has been well-studied by Neugebauer and which, as shown by Rochberg, represent a different scheme to the texts which use *Ziqpu* Stars (Rochberg 2004).

Two basic forms of the rising times scheme are known from Babylonian sources: a calendar-based scheme and a zodiac-based scheme. The calendar-based scheme gives positions at or behind *Ziqpu* Stars which culminate at sunrise and/or sunset on days in the schematic 360-day year. Three texts present this information for the fifteenth day of each month whilst one gives this information for each day of the year. The zodiac-based scheme gives positions at or behind *Ziqpu* Stars which culminate at the moment when a given position within the zodiac rises across the eastern horizon. The texts containing the zodiac-based scheme fall into two main groups: one group contains statements of ranges of positions relative to *Ziqpu* Stars which culminate corresponding to the rising of a whole sign of the zodiac across the eastern horizon; the second group contains statements of the position which culminates corresponding to the rising of the end of a so-called microzodiac sign (1/12 of a sign of the zodiac). Within this latter group are texts which attest to a standard series which covered the whole twelve signs of the zodiac beginning with the middle of Aries and which sets out the rising time scheme along with some material utilizing the microzodiac signs to make astrological associations with stars which "expel a flare", and simplified texts which present the rising time scheme on its own.

6   1   Introduction

Table 1.2 Sources containing material concerning rising time schemes

Tablet	Type	Previous publication
A 3414 + U 181a, b, c, d	Day scheme	Schaumberger (1955)
A 3427	Microzodiac scheme	Schaumberger (1955), Rochberg (2004)
BM 32276	Microzodiac scheme; related material	Unpublished
BM 34639 (= LBAT 1501) Obv. II	Month scheme	Unpublished
BM 34664 (= LBAT 1503)	Microzodiac scheme	Rochberg (2004)
BM 34713 (= LBAT 1499) Rev. I 10ff.	Microzodiac scheme	Rochberg (2004)
BM 36609+	Zodiac scheme; related material	Roughton et al. (2004).
BM 38704	Month scheme	Steele (2014) [without discussion]
BM 45456 (= LBAT 1505)	Simplified microzodiac scheme	Unpublished
BM 46167	Simplified microzodiac scheme	Unpublished
BM 77242	Simplified microzodiac scheme	Horowitz (1994) [without discussion]
BM 78161 (The "GU Text")	Related material	Pingree and Walker (1988)
W 22281a (= SpTU I 95)	Month scheme	Hunger (1976) [without discussion]

References

Horowitz W (1994) Two New Ziqpu-Star Texts and Stellar Circles. Journal of Cuneiform Studies 46:89–98
Hunger H (1976) Spätbabylonische Texte aus Uruk. Teil I. Gebr. Mann, Berlin
Hunger H, Pingree D (1989) MUL.APIN. An Astronomical Compendium in Cuneiform, Archiv für Orientforschung Beiheft 24. Berger & Söhne, Horn
Montelle C (2016) The *Anaphoricus* of Hypsicles of Alexandria. In: Steele JM (ed) The circulation of astronomical knowledge in the ancient world. Brill, Leiden, pp 287–315
Neugebauer O (1936) Jahreszeiten und Tageslängen in der babylonischen Astronomie. Osiris 2:517–550
Neugebauer O (1953) The rising-times in Babylonian astronomy. J Cuneif Stud 7:100–102
Neugebauer O (1975) A history of ancient mathematical astronomy. Springer, Berlin
Pingree D, Walker CBF (1988) A Babylonian star catalogue: BM 78161. In: Leichty E et al (eds) A scientific humanist: studies in memory of Abraham Sachs. University Museum, Philadelphia, pp 313–322
Rochberg F (2004) A babylonian rising time scheme in non-tabular astronomical texts. In: Burnett C, Hogendijk JP, Plofker K, Yano M (eds) Studies in the history of the exact sciences in honour of David Pingree. Brill, Leiden, pp 56–94
Roughton NA, Steele JM, Walker CBF (2004) A Late Babylonian Normal and *Ziqpu* Star text. Arch Hist Exact Sci 58:537–572

Schaumberger J (1955) Anaphora und Aufgangskalender in neuen Ziqpu-Texten. Zietschrift für Assyriologie 52:237–251

Steele JM (2013) Shadow-length schemes in babylonian astronomy. SCIAMVS 14:3–39

Steele JM (2014) Late Babylonian *Ziqpu*-Star lists: Written or remembered traditions of knowledge? In: Bawanypeck D, Imhausen A (eds) Traditions of written knowledge in ancient Egypt and Mesopotamia, Alter Orient und Altes Testament 403. Ugarit-Verlag, Münster, pp 123–151

Chapter 2
Preliminaries

Abstract Babylonian rising time schemes rely upon several concepts and techniques of Babylonian astronomy. This chapter outlines the Babylonian calendar (in particular the schematic calendar of 360 days), the tradition of schematic astronomy based upon the early astronomical work known as MUL.APIN, the use of the culmination of certain stars (known as *Ziqpu* Stars) to indicate specific moments of time, and the Babylonian development of the zodiac, a mathematical division of the ecliptic into twelve equal parts.

Keywords Babylonian astronomy · Calendar · Culminating point · MUL.APIN · Schematic astronomy · *Ziqpu* stars · Zodiac

2.1 The Calendar

The Babylonian calendar was a luni-solar calendar. The beginning of a month was determined by whether the new moon crescent was visible on the thirtieth evening of the previous month. If the moon was visible then the day just beginning was renamed the first day of the new month, the previous month then having 29 days; if the moon was not visible, then the day just beginning was confirmed as the thirtieth day of the current month and the new month would begin on the following evening. Various methods for calculating in advance whether the new moon crescent would be visible on the thirtieth evening were developed by the Babylonian astronomers during the first millennium (Brack-Bernsen 2002). There is considerable evidence that by about 300 BC, and perhaps earlier, month lengths were usually determined in advance by calculation rather than by means of observation (Steele 2007a). Twelve lunar months last about 10 days short of a solar year. In order to keep months roughly in line with the seasons, the Babylonians intercalated either a second Month VI or a second Month XII every 3 years or so. By the fifth century BC, a regular cycle of seven intercalations every 19 years had been adopted (Britton 2007).

© The Author(s) 2017
J.M. Steele, *Rising Time Schemes in Babylonian Astronomy*, SpringerBriefs
in History of Science and Technology, DOI 10.1007/978-3-319-55221-7_2

The fact that in a luni-solar calendar a month can be either 29 or 30 days long and that a year can contain either 12 or 13 months can make it inconvenient to use in calculations. This problem was avoided as early as the third millennium BC in Mesopotamia by the use of schematic calendars which set each month as containing 30 days and, at least from the early second millennium BC onwards, each year as containing exactly 12 months (i.e. without intercalation), making a total of 360 days in a year (Brack-Bernsen 2007). This schematic calendar was often used to simplify calculations in administrative contexts, such as the calculation of work rates, interest, etc., and was adopted in astronomy already in the Old Babylonian period. In the first millennium BC, a branch of astronomy which I term "schematic astronomy", which developed out of the important early astronomical work MUL. APIN (see Sect. 2.2), relied upon the notion of the schematic calendar of twelve thirty-day months with the solstices and equinoxes placed on day 15 of Months I, IV, VII and X. It is important to note that the schematic calendar never operated as a real-world calendar: it only acted either in a completely schematic framework, unconnected to the actual calendar, or as a computational convenience, allowing calculations to be made that could then be mapped onto the actual luni-solar calendar.

2.2   Schematic Astronomy

I use the term "schematic astronomy" to refer to a branch of astronomy in the first millennium BC which developed out of the astronomical tradition represented by second or early first millennium BC astronomical texts including tablet 14 of *Enūma Anu Enlil* (Al-Rawi and George 1992), the Three-Stars-Each texts (Horowitz 2014), and MUL.APIN (Hunger and Pingree 1989). These early texts all rely upon the schematic 360-day calendar and use simple monthly zigzag functions to model astronomical phenomena such as the variable length of daylight during the year and the duration of visibility of the moon through a month. Zigzag functions are simple mathematical computational tools where a function varies between a maximum and a minimum value in a series of steps of constant difference. For example, in MUL.APIN II ii 43–iii 12, the length of night is said to correspond to 3 minas weight of water in a waterclock on the 15th day of Month I, to 2 2/3 minas of water on the 15th of Month II, to 2 1/3 minas on the 15th of Month III, and to 2 minas on the 15th of Month IV. Each month the weight of water, which is assumed to be directly proportional to the passage of time, decreases by 1/3 mina. After Month IV, the length of night increases by 1/3 mina per month up to a maximum of 4 minas in Month X. Thus, on the 15th of Month V, night lasts 2 1/3 minas, on the 15th of Month VI it lasts 2 2/3 minas, etc. After Month X, the length of night decreases again by the same difference of 1/3 mina per month until we get back to Month I where we again find that night equals 3 minas of water. In later Babylonian mathematical astronomy, zigzag functions with non-integer periods were used (i.e. sequences which do not return to the starting value after one cycle through the maximum and minimum, but only after many cycles), but only simple zigzag functions with integer periods (usually either 12 months or 30 days) appear in texts of schematic astronomy.

During the Late Babylonian period, schematic astronomy co-existed alongside observational astronomy, goal-year astronomy (a type of astronomy in which predictions of future astronomical phenomena were made by applying lunar and planetary periods to past observations), and mathematical astronomy. Although to our eyes the schematic astronomy tradition seems primitive, largely divorced from empirical reality and fundamentally incompatible with these other branches of Late Babylonian astronomy, it does not seem that for the Babylonians schematic astronomy was considered any less a part of astronomy. Indeed, it is almost certain that the scribes who wrote texts containing schematic astronomy were the same scribes who recorded observations and made calculations using mathematical astronomy.

Fundamental to all schematic astronomy in the Late Babylonian period seems to have been the classic text of early Babylonian astronomy known as MUL.APIN. MUL.APIN is a compendium of astronomical and astrological material that was put together sometime in the late second or, in my opinion more likely, the early first millennium BC. It is preserved in many copies from both Assyria and Babylonia, including copies written as late as the last few centuries BC. The preserved copies attest to a very stable text, albeit with flexibility in layout on tablets. The modern edition reconstructs the text as written in a two-tablet series, and that is how I will cite the text. Tablet 1 of MUL.APIN contains a series of lists of stars[1]: (1) three lists of stars in the paths of Enlil, Anu and Ea, which correspond to stars within northerly, central and southerly ranges of declination; (2) a list of the dates in the schematic calendar of the first appearances (heliacal rising) of stars; (3) a list of simultaneously rising and setting stars; (4) a list of the number of days between the first appearances of stars; (5) a list of *Ziqpu* Stars which culminate in order (see Sect. 2.3); (6) a list of dates in the schematic calendar when stars culminate as other stars rise; and (7) a list of stars which stand in the path of the moon (i.e., zodiacal constellations). Tablet 2 contains a more diverse range of material including statements that the sun and the five planets pass through the same stars as the moon, procedures for determining intercalation, statements about the subdivision of the synodic arcs of each of the planets, a mathematical scheme for the length of shadow cast at different times of day on the days of the solstices and equinoxes, a mathematical scheme for the length of daylight and the duration of visibility of the moon, and a collection of celestial omens.

---

[1]No distinction is made between "star" and "constellation" in Akkadian; many of the "stars" I will refer to are groups of stars which represent a part or the whole of a constellation.

Several key concepts, methods and parameters from MUL.APIN underlie the Late Babylonian tradition of schematic astronomy including:

1. The schematic 360-day calendar with the solstices and equinoxes placed on the 15th day of Months I, IV, VII and X.
2. The length of daylight modelled as a zigzag function for the 1st and 15th days of each month with extrema in the ratio of 2:1.
3. The lunar visibility modelled as a zigzag function with the duration of visibility taken to be zero on the 30th day of a month and the whole night on the 15th day of the month. The daily change in the duration of lunar visibility is therefore equal to 1/15 of the length of night on the 15th of the month.
4. The length of a shadow cast by a gnomon multiplied by the time since sunrise is equal to a constant. The value of the constant on the 15th of each month is given by a zigzag function with extrema in the ratio 3:2.

A number of previously published texts can be seen to draw upon these foundations. For example, several late texts concerning shadlow-length schemes can be seen as expansions of items 1 and 4 (Steele 2013), one section of a Seleucid period compendium of material dealing with the calculation of month lengths and lunar visibilities (among other things) is founded upon the principles of items 2 and 3 (TU 11 Sect. 19; see Brack-Bernsen and Hunger 2002), and a late "reworking" of MUL.APIN draws upon all of these foundations.[2] I will argue in Chap. 6 that the rising time schemes draw upon items 1 and 2 of this list and should be understood as another example of schematic astronomy. The rising time scheme is particularly interesting in this regard because it demonstrates that new astronomical concepts, such as the zodiac, were incorporated within the tradition of schematic astronomy.

Finally, it is worth remarking that whereas most Late Babylonian astronomical texts contain either *data* relating to specific events (whether they be observed, predicted using Goal-Year methods, or calculated using mathematical astronomy) or *procedures* written in the second person instructing the reader how to make calculations or predictions, the schematic astronomical texts are generally *descriptive*, written in the third person and presenting a set of astronomical facts. These facts are independent of time (except in the sense of modelling variations over the year)—they are assumed to be a reflection of how the universe is in general rather than how it appears on a given occasion. Furthermore, no procedures are given for how these facts could be used in calculations. Thus, I suggest that the purpose of the schematic astronomy texts was to provide a theoretical mathematical description of celestial phenomena. I will return to the question of the purpose of schematic astronomy in Chap. 6.

---

[2]I have so-far identified five fragments of copies of this late reworking and will publish a full study of this work in due course.

2.3   *Ziqpu* Stars

The Akkadian word *ziqpu* is used in astronomical contexts to refer to the highest point ("culmination") of a heavenly body. Due to the daily rotation of the Earth, as viewed from a given location in the northern hemisphere, most stars appear to rise in the east, rotate across the night sky, crossing the meridian at their highest point, before setting in the west. Only the circumpolar stars deviate from this behavior, circling around the north celestial pole without reaching low enough to set below the horizon. Nevertheless, even the circumpolar stars reach their highest (and their lowest) point when crossing the meridian. Ignoring precession and the proper motion of stars, concepts which were unknown to the Babylonians, the stars remain in a fixed relationship to one another. As a result, the time interval between the culminations of two given stars is always the same. This property makes the culmination of stars a useful tool for measuring the passage of time at night. Unsurprisingly, therefore, Babylonian scholars made use of culminating stars along with water clocks to measure time during the hours of darkness.

The use of culminating stars to keep track of time can be traced back to at least the early first millennium BC, but may very well predate this period (Steele 2014). MUL.APIN contains a list of fourteen constellations whose culmination should be observed. Interestingly, this group of constellations is referred using a name which refers directly to their use: MUL.MEŠ *šá ziq-pi* "*Ziqpu* Stars". In the Neo-Assyrian period, we find evidence that the *Ziqpu* Stars were used in a wide range of contexts: a letter referring to a storm during the night notes that the beginning and the end of the storm occurred when certain *Ziqpu* Stars culminated, a ritual text indicates that certain parts of the ritual are to be performed when specific *Ziqpu* Stars culminate, two reports of observations of lunar eclipses time the event using *Ziqpu* Stars, and a collection of blessings from Ḫuzirina mentions the *Ziqpu* Stars as a group.[3] In addition, a small fragment which parallels a completely preserved text from much later Seleucid Uruk contains a list of distances between *Ziqpu* Stars.

Ziqpu Stars appear in a variety of contexts in the Late Babylonian period. For example, astrological texts relate the culmination of *Ziqpu* Stars with predictions for the life of an individual at birth,[4] and several fragmentary texts of schematic astronomy mention *Ziqpu* Stars.[5] In the observational texts, records of eclipses sometimes include a statement about when the eclipse began relative to the culmination of a *Ziqpu* Star. The report of the eclipse of 2 August 123 BC recorded in an Astronomical Diary is typical:

---

[3]For the report of the storm, see Lanfranchi and Parpola (1990: No. 249); for the ritual text, see van Driel (1969: 90–93); for the eclipse reports, see Parpola (1993: No. 134 and 139); for the blessing, see Horowitz (1994: 97).

[4]TU 14 and its duplicate U 197; see Sachs (1952).

[5]For example, BM 65756 (unpublished).

5 UŠ *ár* MÚL DELE *ziq-pi sin* AN.KU$_{10}$
(When the point) 5 UŠ behind the Single Star culminated, lunar eclipse.[6]

Unlike the Neo-Assyrian examples cited above, this and several other of the Late Babylonian eclipse report refer not just to the culmination of a *Ziqpu* Star but to the culmination of a point a distance behind (or, more rarely, in front of) the *Ziqpu* Star. This distance is recorded with the unit UŠ, which may be translated as "degree". In modern terms, this distance corresponds to the difference in Right Ascension between the star and the point which is culminating. Because it takes one day for the sky to perform one complete revolution, and because the Babylonians measured time using the same unit UŠ, this difference in Right Ascension can be equated with the time difference between the star and the point culminating. Thus, 1 UŠ = 1° of celestial rotation = 4 min of time. The rising time schemes use the same convention of stating positions behind *Ziqpu* Stars given in UŠ and the related units NINDA (1 UŠ = 60 NINDA) and *bēru* (1 *bēru* = 30 UŠ).

More than a dozen lists of *Ziqpu* Stars are known from sites throughout Babylonia (Steele 2014). These lists exist in two main forms: lists that give the distances in *bēru* and UŠ between consecutive *Ziqpu* Stars, and lists that give other information instead of these distances. The lists attest to a stable core repertoire of 25 *Ziqpu* Stars. Almost all of the *Ziqpu* Stars attested in other texts (astrological texts, observational texts, rising time texts) are taken from this core list of 25 stars. Those that are not are few and are always part of a constellation that already contains *Ziqpu* Stars (for example, the 25 star list includes 4 *Ziqpu* Stars in the constellation The Panther[7]: The Shoulder of the Panther, The Bright Star of its Chest, The Knee, and The Heel. The rising time text SpTU I 95 adds a fifth *Ziqpu* Star in this constellation, The Feet of the Panther). A more significant case is with the constellation The Twins. One *Ziqpu* Star list (AO 6478 and its duplicate K.9794) includes an entry for a Rear Twins in addition to a (Front) Twins Star, separated by 5 UŠ. The Front and Rear Twins both appear in the rising time texts.

Although several tablets containing *Ziqpu*-Star lists are known which include statements of the distances between the *Ziqpu* Stars, all but one tablet is damaged and do not preserve all of the entries.[8] Unfortunately, the fully preserved list, AO 6478, differs from all of the other lists both in style (all other lists have entries in the form *x ana* SN "*x* to SN", where *x* is a distance and SN is a star name, whereas entries on AO 6478 have the form TA SN$_1$ EN SN$_2$ *x* "From SN$_1$ to SN$_2$ *x*"), and in

---

[6]Diary No. -122D Obv. 8 (my translation).

[7]I follow Hunger and Pingree (1989) in translating MÚL.UD.KA.DUḪ.A as "The Panther" (Akkadian *nimru*), based upon the arguments presented in Parpola (1983: 93). Literally, MÚL.UD. KA.DUḪ.A in Sumerian is "Demon with the Gaping Mouth" and the Seleucid period uranology text MLC 1866 describes this constellation as a human figure. It is unclear whether MLC 1866 reflects a late re-interpretation of the name or whether the constellation had always been seen as a human figure rather than a panther. In the absence of compelling evidence either way, I translate "panther" simply to maintain consistency with earlier publications.

[8]See Table 2 in Steele (2014) for a summary of the preserved entries in the various lists.

the number of stars included in the list (AO 6478 is the only list to include the Rear Twins). Furthermore, the total of the distances given in AO 6478 is 364 UŠ, whereas other sources such as the Neo-Assyrian blessing referred to above explicitly refer to the total circuit of the *Ziqpu* Stars as being 360 UŠ,[9] which is what we would expect. It may well be that the text of AO 6478 is corrupt in giving the total as 364 UŠ because the entry for the Rear Twins Star is given in a different format to the other entries in the list: it reads *bi-rit* MUL.MAŠ.TAB.BA 5 UŠ "between the Twins 5 UŠ", rather than the expected TA MUL.MAŠ.TAB.BA EN MÚL.MAŠ.TAB.BA EGIR-*i* 5 UŠ "From the (Front) Twin to the Rear Twin 5 UŠ".[10] It is possible that the author of AO 6478 intended that this should be interpreted as meaning that the Twins constellation extended for 5 UŠ, and that the interval between the Twins and the next star in the list was to be taken as being from the Twins, skipping over the distance between the two Twins. This interpretation brings the entries into agreement with the 25 star *Ziqpu* Star lists which omit the Rear Twin but have the same value for the distance between the Twins and the next star. The rising time texts bring some clarity to this problem. As will be discussed in the commentary to BM 35456 (Sect. 4.3.1), it is now apparent that the distance of 30 UŠ between The Handle of the Crook and The Twins given in the *Ziqpu* Star lists is to be understood as the distance between The Handle of the Crook and The Rear Twin, with the Front Twin placed 5 UŠ in front of The Rear Twin and therefore 25 UŠ behind The Handle of the Crook.

There is one further difficulty in establishing the *Ziqpu* Star list, however. Even with the new understanding of the positions of the two Twins Stars, the total distance for the circuit of the *Ziqpu* Stars implied by AO 6478 would still not be the expected 360 UŠ, but rather 359 UŠ. So far as they are preserved, the distances between the *Ziqpu* Stars in all of the other lists are identical with those found on AO 6478. However, none of the other lists preserve all of the distances in the list and even combining all of the sources, there remains a gap in the entries around the stars the Yoke and the Rear Harness.[11] The sequence of stars in this part of the list is Eru, The Harness, The Yoke, The Rear Harness and The Circle.[12] The distances between these stars in AO 6478 are 25 UŠ between Eru and The Harness, 8 UŠ between The Harness and The Yoke, 9 UŠ between The Yoke and The Rear Harness and 12 UŠ between The Rear Harness and The Circle, totaling 54 UŠ. As I will discuss in Sect. 4.5.1, the rising time texts imply a distance of 55 UŠ between Eru and The Circle. It is worth noting also that this part of the list on AO 6478 is the only part

---

[9]The text (STT II 340, Obv. 12) reads: 12 DANNA MUL.MEŠ [*z*]*iq-pi šá* KASKAL *šu-ut* ᵈ*en-lil*"12 *bēru* are the *Ziqpu* Stars in the path of Enlil". 1 *bēru* = 30 UŠ therefore 12 *bēru* = 360 UŠ.

[10]See already Koch (1996).

[11]The only list to preserve these entries is the Sippar planisphere but the text is very badly damaged at this point and Horowitz and Al-Rawi's (2001) readings of the traces are clearly based upon what they expected to find using AO 6478 as a model.

[12]It is worth noting here that the list VAT 16436, which does not contain values for the distances in UŠ, mixes up the order of the entries in this part of the list, swapping the Harness and the Rear Harness.

Table 2.1 The 25 core *Ziqpu* Stars. Uncertain distances are given in parentheses (note that while these three individual distances may be uncertain, their total is not)

Star	Distance to next star (UŠ)	Additional stars
The Shoulder of the Panther	10	
The Bright Star of its Chest	20	
The Knee	20	10 UŠ to The Foot of the Panther
The Heel	10	
The Four Stars of the Stag	15	
The Dusky Stars	15	
The Bright Star of the Old Man	10	
Nasrapu	15	
The Crook	10	
The Handle of the Crook	30	25 UŠ to The Front Twin
The Twins (The Rear Twin)	20	5 UŠ behind The Front Twin
The Crab	20	5 UŠ to The Rear Stars of the Crab
The 2 Stars of the Head of the Lion	10	
The 4 Stars of its Breast	20	
The 2 Stars of its Thigh	10	
The Single Star of its Tail	10	
Eru	25	
The Harness	(10)	
The Yoke	(10)	
The Rear Harness	(10)	
The Circle	15	
The Star from the Doublets	5	
The Star from the Triplets	10	
The Single Star	10	
The Lade of Life	20	

where the distances are not all multiples of 5 UŠ. I tentatively suggest either that the distances between The Harness and The Yoke, between The Yoke and The Rear Harness and between the Rear Harness and The Circle were assumed to be each 10 UŠ or that the distances between The Harness and The Yoke and between The Rear Harness and The Circle were 8 UŠ and 12 UŠ as in AO 6478, but that the distance between The Yoke and The Rear Harness be increased to 10 UŠ. Without further evidence, I assume the former (simpler) option.

Table 2.1 lists the 25 core *Ziqpu* Stars together with the distances between then in UŠ, so far as these can be established on the basis of the various *Ziqpu*-Star lists and the assumptions made above. In addition, I give the positions of the additional stars not in the 25-star list which appear in the rising time texts. It is worth noting that the preserved lists do not all begin with the same star. For this reason, I have chosen to arbitrarily begin the list with The Shoulder of the Panther. This star provides the starting point for the rising time schemes. Finally, it is worth noting that the preserved lists exhibit considerable variation in the writing of star names. I have elsewhere argued that this may be because the list was something that was remembered by the scribes and occasionally written down rather than being a text which was copied (Steele 2014).

Two ways of referring to the culmination of *Ziqpu* Stars are found in the rising time texts: (1) the expression *ana ziq-pi* DU, literally meaning "goes to its highest point", immediately following the name of the *Ziqpu* Star; and (2) the expression *ina* UGU or *ina muḫ-ḫi*, literally "in its topmost" immediately before the star name.[13] The two expressions do not appear to have any functional difference and so both can be translated simply as "culminate". In order to maintain a distinction which reflects the difference in syntax of the original texts, I translate the first expression "(star name) culminates" and the second "at the culmination of (star name)".

2.4 The Zodiac

The development of the zodiac as a uniform division of the path of the sun, moon and planets was a key step in Babylonian astronomy. Motion through the zodiac may be tracked through its division into twelve equal parts, each of which is subdivided into 30 UŠ. The position of a body within a zodiacal sign is its celestial longitude and since the complete zodiac contains $12 \times 30 = 360$ UŠ, we can translate UŠ as degree if we wish. Positions perpendicular to this motion, broadly equivalent to modern celestial longitudes, can be measured above or below the middle of the zodiacal band (Steele 2007b).

The zodiac has its origin in two concepts within earlier Babylonian astronomy: a list of the constellations which stand in the path of the moon (in modern terms, the zodiacal constellations) and the schematic calendar. MUL.APIN contains a list of eighteen zodiacal constellations. Observational texts from before the beginning of the 4th century BC often report the position of the moon or a planet either within, in front of, or behind one of these constellations. Sometime towards the end of the fifth century BC,[14] the model of the schematic calendar in which the year contains twelve 30-day months was used to create an equivalent "schematic" zodiac in which there were twelve constellations each of which contain 30 UŠ. Just as the schematic

---

[13]In other contexts, *ina* UGU is often used simply to add emphasis to an expression and can be translated simply as 'at', and such a translation was, for example, used by Lanfranci and Parpola (1990: 178) in their edition of the letter referring to the night-time storm discussed above. For further discussion of why I believe it is correct to translate this phrase as 'culminate' when the text is referring to *Ziqpu* stars, see Steele (2014).

[14]Possible dates for the development of the zodiac have been proposed for between the middle of the fifth century BC (Rochberg-Halton 1991) to within a few years of 400 BC (Britton 2010).

calendar simplified mathematical calculation, the uniformly divided zodiac simplified astronomical calculation. A useful consequence of the parallelism of the schematic calendar and the zodiac is that in one day the sun's mean motion will be equal
to 1 UŠ.[15] Assuming the sun is at the beginning of the zodiac (0° in Aries) at the
beginning of the solar year, then the schematic date and the sun's position are equal.
For example, on Month V day 10, the sun will be 10° within the fifth sign of the
zodiac (Leo). A consequence of this is that when using the schematic calendar the
solstices and equinoxes are placed at 15° within Aries, Cancer, Libra and Capricorn.
It should be noted, however, that the zodiac was not only used in the context of the
schematic calendar. For example, in the System A and System B lunar theories,
which operate with lunar months and a value for the true length of the solar year, the
vernal equinox is placed at 10° and 8° within Aries respectively.

The signs of the zodiac were named after zodiacal constellations that fell within
the relevant section of the path of the moon. Some texts preserve two alternative
names for two of the signs, presumably because the standard names had not yet
been agreed upon. In my translations I use the modern name of the zodiacal sign
when the text is referring to a sign and give a literal translation of the name when it
is used as a constellation. For example, I translate MÚL.ALLA as "Cancer" when
used as a zodiacal sign but as "The Crab" when used as a constellation name.
Similarly, I render MÚL.ḪUN as "Aries" when it is a zodiacal sign but "The Hired
Man" when it is a constellation.

Some of the rising time texts make use of a division of each zodiacal sign into
twelve "micro-signs", each of 2½°. The micro-signs are named after the regular
signs and given in sequence beginning with the same sign as the governing zodiacal
sign. Thus, the first microzodiac sign of Aries is Aries, the second micro-zodiacal
sign of Aries is Taurus, etc., up to the twelfth microzodiac sign of Aries which is
Pisces. In Taurus, the first microzodiac sign is Taurus, the second is Gemini, etc.,
up to the twelfth microzodiac sign of Taurus which is Aries. The full system of
microzodiac signs is shown in Table 2.2. In the rising times texts, the microzodiac
signs are referred to both by their number within the governing zodiacal sign and by
their microzodiac sign name given together with the governing zodiacal sign name.
For example:

9-tú ḪA.LA šá MÚL.RÍN MÚL.MAŠ šá MÚL.RÍN
9th portion of Libra (which) is Gemini of Libra

Outside of the rising time texts, the system of micro-zodiacal signs is found in
various astrological texts, most notably the so-called "microzodiac series" which
correlates various items including *materia medica* and hemerological material with
the micro-zodiacal and zodiacal signs.[16]

---

[15]The parallelism between schematic dates and the sun's mean motion produces a value of 13 UŠ
per day for the mean motion of the moon, a value which underlies the so-called *Dodecatemoria*
and *Kalendertext* astrological schemes. See Brack-Bernsen and Steele (2004).
[16]Weidner (1967), Monroe (2016).

Table 2.2 The micro-zodiac system

Portion	Aries	Taurus	Gemini	Cancer	Leo	Virgo	Libra	Scorpio	Sagittarius	Capricorn	Aquarius	Pisces
1	Aries	Taurus	Gemini	Cancer	Leo	Virgo	Libra	Scorpio	Sagittarius	Capricorn	Aquarius	Pisces
2	Taurus	Gemini	Cancer	Leo	Virgo	Libra	Scorpio	Sagittarius	Capricorn	Aquarius	Pisces	Aries
3	Gemini	Cancer	Leo	Virgo	Libra	Scorpio	Sagittarius	Capricorn	Aquarius	Pisces	Aries	Taurus
4	Cancer	Leo	Virgo	Libra	Scorpio	Sagittarius	Capricorn	Aquarius	Pisces	Aries	Taurus	Gemini
5	Leo	Virgo	Libra	Scorpio	Sagittarius	Capricorn	Aquarius	Pisces	Aries	Taurus	Gemini	Cancer
6	Virgo	Libra	Scorpio	Sagittarius	Capricorn	Aquarius	Pisces	Aries	Taurus	Gemini	Cancer	Leo
7	Libra	Scorpio	Sagittarius	Capricorn	Aquarius	Pisces	Aries	Taurus	Gemini	Cancer	Leo	Virgo
8	Scorpio	Sagittarius	Capricorn	Aquarius	Pisces	Aries	Taurus	Gemini	Cancer	Leo	Virgo	Libra
9	Sagittarius	Capricorn	Aquarius	Pisces	Aries	Taurus	Gemini	Cancer	Leo	Virgo	Libra	Scorpio
10	Capricorn	Aquarius	Pisces	Aries	Taurus	Gemini	Cancer	Leo	Virgo	Libra	Scorpio	Sagittarius
11	Aquarius	Pisces	Aries	Taurus	Gemini	Cancer	Leo	Virgo	Libra	Scorpio	Sagittarius	Capricorn
12	Pisces	Aries	Taurus	Gemini	Cancer	Leo	Virgo	Libra	Scorpio	Sagittarius	Capricorn	Aquarius

References

Al-Rawi F, George A (1991–1992) Enūma Anu Enlil XIV and other early astronomical tables.
   Archiv für Orientforschung, pp 38–39, 52–73
Brack-Bernsen L (2002) Predictions of lunar phenomena in Babylonian astronomy. In: Steele JM,
   Imhausen A (eds) Under one sky: astronomy and mathematics in the ancient near east.
   Ugarit-Verlag, Münster, pp 5–19
Brack-Bernsen L (2007) The 360-day year in Mesopotamia. In: Steele JM (ed) Calendars and
   years: astronomy and time in the ancient near east. Oxbow Books, Oxford, pp 83–100
Brack-Bernsen L, Hunger H (2002) TU11: a collection of rules for the prediction of month lengths.
   SCIAMVS 3:3–90
Brack-Bernsen L, Steele JM (2004) Babylonian mathemagics: two mathematical
   astronomical-astrological texts. In: Burnett C, Hogendijk JP, Plofker K, Yano M
   (eds) Studies in the history of the exact sciences in honour of David Pingree. Brill, Leiden,
   pp 95–125
Britton JP (2007) Calendars, intercalations and year-lengths in Mesopotamian astronomy. In:
   Steele JM (ed) Calendars and years: astronomy and time in the ancient Near East. Oxbow
   Books, Oxford, pp 115–132
Britton JP (2010) Studies in babylonian lunar theory: part iii. the introduction of the uniform
   zodiac. Arch Hist Exact Sci 64:617–663
van Driel G (1969) The cult of aššur. Koninklijke Van Gorcum & Comp, Assen
Horowitz W (2014) The three stars each: the astrolabes and related texts, Archiv für
   Orientforschung beiheft 33. Berger & Söhne, Horn
Horowitz W, Al-Rawi FNH (2001) Tablets from the Sippar library IX. A ziqpu-star planisphere.
   Iraq 63:171–181
Hunger H, Pingree D (1989) MUL.APIN. An astronomical compendium in cuneiform, Archiv für
   Orientforschung beiheft 24. Berger & Söhne, Horn
Koch J (1996) AO 6478, MUL.APIN und das 364 Tage-Jahr, *NABU* 1996/111
Lanfranci GB, Parpola S (1990) The correspondence of Sargon II, part II. Helsinki University
   Press, Helsinki
Monroe MW (2016) Advice from the stars: the micro-zodiac in Seleucid Babylonia. PhD
   dissertation, Brown University
Parpola S (1983) Letters from Assyrian and Babylonian scholars to the kings Esarhaddon and
   Assurbanipal. Part II: commentary and appendices, Alter Orient und Altes Testament 5/2.
   Butzon & Bercker, Kevelaer
Rochberg-Halton F (1991) Between observation and theory in Babylonian astronomical texts.
   J Near East Stud 50:107–120
Sachs A (1952) Babylonian horoscopes. J Cuneif Stud 6:49–75
Steele JM (2007a) The length of the month in Mesopotamian calendars of the first millennium BC.
   In: Steele JM (ed) Calendars and years: astronomy and time in the ancient near east. Oxbow
   Books, Oxford, pp 133–148
Steele JM (2007b) Celestial measurement in Babylonian astronomy. Ann Sci 64:293–325
Steele JM (2013) Shadow-length schemes in Babylonian astronomy. SCIAMVS 14:3–39
Steele JM (2014) Late Babylonian *ziqpu*-star lists: written or remembered traditions of knowledge?
   In: Bawanypeck D, Imhausen A (eds) Traditions of written knowledge in ancient Egypt and
   Mesopotamia, Alter Orient und Altes Testament 403. Ugarit-Verlag, Münster, pp 123–151
Weidner E (1967) Gestirn-darstellungen auf babylonischen tontafeln. Hermann Böhlaus Nachf,
   Vienna

Chapter 3
Calendar-Based Rising Time Schemes

Abstract This chapter presents a previously unknown form of rising time scheme which is founded upon the calendar. Three cuneiform texts whose contents were not previously understood are shown to contain a rising time scheme which relates culminating points at or behind *ziqpu* stars with sunrise and sunset on day 15 of months in the schematic 360-day calendar. The underlying rising time scheme is fully reconstructed and shown to be founded on the assumption from MUL.APIN that the solstices and equinoxes take place on the 15th day of Months I, IV, VII and X in the schematic calendar. Finally, a previously misunderstood text is shown to contain an extension of the monthly scheme to give the culminating points at sunrise and sunset on every day of the schematic year.

Keywords Babylon · Babylonian astronomy · Calendar · Culminating point · Cuneiform tablet · Uruk · *Ziqpu* stars

3.1 Introduction

A previously unidentified form of rising time scheme which relates dates in the ideal calendar to the culmination of points at or behind *Ziqpu* Stars is found on four tablets. This calendar-based scheme lists the point which culminates at sunset and sunrise on specific dates in the 360-day schematic calendar. The scheme is attested in two forms: a monthly scheme which lists the culmination data for the 15th of each month and a daily scheme which gives the culmination data for each day of the schematic year. As I will demonstrate, the scheme divides the year into two halves from summer solstice to winter solstice and from winter solstice to summer solstice, placing the solstice on the 15th of Months IV and X following the tradition of MUL.APIN.

© The Author(s) 2017
J.M. Steele, *Rising Time Schemes in Babylonian Astronomy*, SpringerBriefs in History of Science and Technology, DOI 10.1007/978-3-319-55221-7_3

3.2 Monthly Schemes

Two types of monthly rising time schemes are known: a simple scheme which
provides a straightforward presentation of the positions relative to *Ziqpu* Stars
which culminate at sunset and sunrise on the fifteenth of each month, and an
extended scheme which gives the same positions relative to *Ziqpu* Stars which
culminate at sunrise on the 15th of each month supplemented by additional stars
which are said to be "in balance" with that star. The simple scheme is preserved on
two tablets: BM 34639 and BM 38704. Both tablets are compendium texts combining *Ziqpu* Star lists with the rising time text and, in BM 34639, material relating
to Normal Stars and the moon's motion in latitude. Aside from insignificant
orthographic differences, the rising time text is identical on the two tablets. The
extended scheme is found on a tablet from Uruk, W 22281a (=SpTU I 95).

3.2.1 BM 34639

BM 34639 (Sp II 122), almost certainly from Babylon, is a substantial fragment
from a multi-column tablet (Fig. 3.1). A copy of the tablet by Pinches is published
as LBAT 1501. No edges are preserved, but from context very little is lost from the
left and bottom edges. From the curvature I estimate that about half of the height

Fig. 3.1 BM 34639 obverse

and a little less than half of the width of the tablet remains. Since the preserved fragment contains two columns on each of the obverse and reverse, when complete the tablet was probably divided into three columns on each side.

BM 34639 is a multi-text or compendium tablet. The first column and probably the missing upper part of the second column on the obverse contained a list of *Ziqpu* Stars with distances between the stars (Steele 2014). The preserved part of column II contains the text relating to the rising times. The topic of the missing third column on the obverse is not known. On the reverse, the first preserved column (and almost certainly the now lost column to the right) contains a copy of a text which describes the upper and lower limits of the path of the moon relative to the Normal Stars (a group of stars spaced irregularly through the zodiacal constellations that are used as reference stars for tracking the motion of the moon and the planets) (Steele 2007). The final column begins with a copy of the end of the lunar latitude section (Sect. 3.1) of the so-called Atypical Text E (BM 41004),[1] and ends with an unidentified section. In the following, I concentrate on the copy of the rising time scheme text preserved in Obverse II.

Transliteration

Obv. II

1'	⌜ina ITU.ŠU UD⌝.[15 ^{d}UTU *ina* UGU *kip-pat* ŠÚ-*ma ina* UGU 4 *šá* MÚL.LU. LIM KUR]
2'	*ina* ITU.IZI UD.⌜15⌝ [^{d}UTU *ina* UGU MÚL *šá tak-šá-a-tum* ŠÚ-*ma ina* UGU MÚL.*na-aṣ-ra-pi* KUR]
3'	*ina* ITU.KIN UD.15 $^{rd⌝}$[UTU MÚL.GAŠAN.TIN ŠÚ-*ma*]
4'	*ina* UGU ½ DANNA [*ár rit-tu$_4$* GÀM KUR]
5'	*ina* ITU.DU$_6$ ⌜UD⌝.15 ^{d}UTU ⌜*ina*⌝ [UGU MÚL *ku-mar šá* MÚL.UD.KA. DUḪ.A ŠÚ-*ma*]
6'	*ina* UGU 5 UŠ *ár* ⌜MÚL⌝.[AL.L.LUL KUR]
7'	*ina* ITU.⌜APIN UD.15 ^{d}UTU *ina* UGU⌝ [10 UŠ *ár* MÚL SA$_4$ *šá* GABA-*šú* ŠÚ-*ma*]
8'	*ina* UGU ⌜½ DANNA *ár* 4⌝ *šá* GABA-⌜*šú*⌝ [KUR]
9'	*ina* ITU.GAN UD.15 ^{d}UTU ⌜*ina* UGU⌝ 10 ⌜UŠ *ár*⌝ [MÚL.*kin-ṣi* ŠÚ-*ma*]
10'	*ina* UGU ½ DANNA *ár* ⌜MÚL.*e$_4$-ru$_6$*⌝ [KUR]
11'	*ina* ITU.AB UD.15 ^{d}UTU *ina* UGU ⌜4⌝ [*šá* MÚL.LU.LIM ŠÚ-*ma*]
12'	*ina muḫ-ḫi kip-pat* [KUR]

[1]BM 41004 and three duplicates (including these lines from BM 34639) are edited and analyzed by Neugebauer and Sachs (1967). For a further discussion of the latitude scheme, see Brack-Bernsen and Hunger (2005–2006). A fourth duplicate is published by Steele (2012) who also discusses the significance of these duplicates. I have recently identified a fifth duplicate on the reverse of the unpublished tablet BM 36175; the obverse of this tablet contains a *Ziqpu* Star list.

13′ *ina* ITU.ZÍZ UD.15 ᵈUTU *ina* ⸢UGU⸣ [MÚL.*na-aṣ-ra-pi* ŠÚ-*ma*]
14′   *ina* UGU ⸢MÚL⸣ [*šá tak-šá-a-tum* KUR]
15′ *ina* ITU.ŠE UD.15 [ᵈUTU *ina* UGU ½ DANNA *ár rit-tu₄* GÀM ŠÚ-*ma*]
16′   *ina* UGU ⸢MÚL⸣.[GAŠAN.TIN KUR]
17′ *ina* ITU.⸢BAR⸣ [UD.15 ᵈUTU *ina* UGU 5 UŠ *ár* MÚL AL.LUL ŠÚ-*ma*]
18′   *ina* ⸢UGU⸣ [MÚL *ku-mar šá* MÚL.UD.KA.DUḪ.A KUR]
19′ ⸢*ina* ITU⸣.[GU₄ UD.15 ᵈUTU *ina* UGU ½ DANNA *ár* 4 *šá* GABA-*šú* ŠÚ-*ma*]

Translation

Obv. II

1′ Month IV, day [15, the Sun sets at the culmination of The Circle and rises at the culmination of The 4 Stars of the Stag.]
2′ Month V, day 15, [the Sun sets at the culmination of The Star of the Triplets and rises at the culmination of Nasrapu.]
3′ Month VI, day 15, [the Sun sets at the culmination of The Lade of Life and]
4′   [rises] at the culmination of ½ *bēru* [behind The Hand of the Crook.]
5′ Month VII, day 15, the Sun [sets] at [the culmination of The Shoulder of the Panther and]
6′   [rises] at the culmination of 5 UŠ behind [The Crab.]
7′ Month VIII, day 15, the Sun [sets] at the culmination of [10 UŠ behind The Bright Star of its Chest and]
8′   [rises] at the culmination of ½ *bēru* behind The 4 Stars of his Breast.
9′ Month IX, day 15, the Sun [sets] at the culmination of 10 UŠ behind [The Knee and]
10′   [rises] at the culmination of ½ UŠ behind Eru.
11′ Month X, day 15, the Sun [sets] at the culmination of The 4 (Stars) [of the Stag and]
12′   [rises] at the culmination of The Circle.
13′ Month XI, day 15, the Sun [sets] at the culmination of [Nasrapu and]
14′   [rises] at the culmination of The Star [of the Triplets.]
15′ Month XII, day 15, [the Sun sets at the culmination of ½ *bēru* behind The Hand of the Crook and]
16′   [rises] at the culmination of [The Lady of Life.]
17′ Month I, [day 15, the Sun sets at the culmination of 5 UŠ behind The Crab and]
18′   [rises] at the culmination of [The Shoulder of the Panther.]
19′ Month [II, day 15, the Sun sets at the culmination of ½ *bēru* behind The 4 Stars of his Breast and]

Critical Apparatus and Philological Notes

6'   Pinches copy has a what looks to be an AB sign marked with a question mark after the 5. Collation of the original and comparison with the duplicate BM 38704 Rev. 3' reveals this sign to be an UŠ as we would expect.

7'   This line is badly damaged. The reading of the first five signs is confirmed by the duplicate BM 38704 Rev. 4'.

8'   The signs ½ DANNA *ár* are badly damaged but confirmed by the duplicate BM 38704 Rev. 5'.

10' This line is badly damaged but the reading proposed here is confirmed by collation and consistent with the duplicate BM 38704 Rev. 6'.

11' Pinches copied a GAR sign at the end of the preserved part of this line. The reading 4 (the beginning of the star name 4 *šá* MUL.LU.LIM) is confirmed by collation.

12' In contrast with the other entries, in this line the logogram UGU is replaced by the syllabic writing *muḫ-ḫi* (note that UGU and *muḫ* are the same sign).

Commentary

This text gives a series of entries for the middle of each month in the schematic 360-day calendar. Following the date, the entries begin by naming the Sun and then give two statements about the culmination of points either at or behind *Ziqpu* Stars. Comparison with the daily rising time text A 3414+ (see Sect. 3.4) indicates that the two statements refer to the rising and the setting of the Sun and I have restored the text appropriately. Restorations of lost star names are based upon the full rising time scheme reconstructed in Sect. 4.5.

The preserved entries from this scheme are summarized in Table 3.1. Considering first the entries for sunrise, by reference to Table 2.1 it is possible to determine the distance in UŠ between the culminating points in successive months. For example, in Month VIII the point ½ *bēru* behind The 4 Stars of his Breast culminated at sunrise and in Month IX the point ½ *bēru* behind Eru culminates at sunrise. From Table 2.1 we find that between The 4 Stars of his Breast and the next *Ziqpu* Star, The 2 Stars of his Thigh, is 20 UŠ. Since we are starting ½ *bēru* (=15 UŠ) behind The 4 Stars of his Breast, it is 5 UŠ to The 2 Stars of his Thigh. From The 2 Stars of his Thigh to Eru is 25 UŠ. Adding a further ½ *bēru* (=15 UŠ) to

Table 3.1 Preserved entries in the simple monthly rising time scheme preserved on BM 34639 and BM 38704

Date	Point culminating at sunset	Point culminating at sunrise
VI 15	[...]	½ *bēru* [behind The Hand of the Crook
VII 15	[...]	5 UŠ behind [The Crab]
VIII 15	[...]	½ *bēru* behind The 4 Stars of his Breast
IX 15	10 UŠ behind [The Knee]	½ *bēru* behind The Frond
X 15	The 4 (Stars) [of the Stag]	The Circle
XI 15	[...]	The Star [of the Triplets]

move beyond Eru brings us to a total of 40 UŠ. The same total of 40 UŠ is found between the entries for Month VI and Month VII, between Month VII and Month VIII, and between Month IX and Month X. Between Month X and Month XI, however, the distance between the culminating points at sunrise is only 20 UŠ. Only two entries preserve the culminating points for sunset: Months IX and X. Here we find a distance of 20 UŠ between the culminating points.

It is significant that the scheme presented in this text gives entries for day 15 of each month, rather than for day 1. In MUL.APIN and the ensuing tradition of schematic astronomy (see Sect. 2.2), the solstices and equinoxes are placed on day 15 of Months I, IV, VII and X. Furthermore, several texts in the schematic astronomy genre such as the shadow-length texts BM 29371, BM 45721 and W 23273 (=SpTU IV 172) begin their entries with the day of the summer solstice on day 15 of Month IV. Although the beginning and end of our text are lost, it seems very likely that the scheme presented here also begins with the date of summer solstice on Month IV Day 15 and cycles through the months ending with Month III Day 15.[2] This fact alone places the monthly rising time scheme within the tradition of schematic astronomy, a conclusion that will be confirmed later in this study.

It is worth noting that the points which culminate when the sun is setting are those that culminate when the sun is rising 6 months later. In the context of the schematic calendar, this rule is astronomically correct since the sun's position in the zodiac will have changed by exactly 180° (assuming constant solar motion) after 6 months (=180 days) or exactly half a year. Over the course of daylight, exactly half of the zodiac, or 180° will rise over the eastern horizon. The cumulative effect, therefore, is that the sun's position at sunrise will be 180° + 180° = 360° or the same at sunrise as it was at sunset six months earlier. This same rule is applied explicitly in the text SpTU I 95 (see Sect. 3.3). As a result we can restore the culminating points at sunrise from those at sunset six months later and vice versa.

3.2.2 BM 38704

BM 38704 (80–11–12, 588), almost certainly from Babylon, is a fragment from the left edge of a tablet (Fig. 3.2). The obverse contains a *Ziqpu* Star list giving the distances between *Ziqpu* Stars while the reverse contains a copy of the monthly rising time scheme text found on BM 34639.[3] This combination of material is identical to that found in the first two columns of the obverse of BM 34639, suggesting that it existed as a single body of text. They are not direct copies of one another, however: BM 38704 uses MUL whereas BM 34639 uses MÚL, and BM

---

[2]On the basis of the preserved entries, we cannot rule out the possibility of the text beginning with Month III Day 15 and ending with Month II Day 15. No other texts are known which follow such an arrangement, however, in contrast to the many which begin in Month IV.

[3]For an edition and discussion of the *Ziqpu* Star list, see Steele (2014).

Fig. 3.2 BM 38704 reverse

38704 gives day numbers in the form UD.15.KAM whereas BM 34639 omits the
final KAM, for example.

The rising time material on the reverse of BM 38704 is a duplicate with only
minor variations to the text on BM 34639 and I therefore only provide a transliteration and critical apparatus here.

Transliteration

1'   [*ina*] ⌜½ DANNA⌝ [*ár rit-tu₄* GÀM KUR]
2'   [*ina*] ⌜ITU⌝.DU₆ ⌜UD⌝.[15.KAM ᵈUTU *ina* UGU MUL.*ku-mar šá* MUL UD.
          KA.DUḪ.A ŠÚ-*ma* …]
3'   *ina* 5 UŠ *á*[*r* MUL.AL.LUL KUR]
4'   *ina* ITU.APIN UD.⌜15⌝.[KAM ᵈUTU *ina* UGU 10 UŠ *ár* MUL.SA₄ *šá*
          GABA-*šú* ŠÚ-*ma*]
5'   *ina* ½ DANNA ⌜*ár*⌝ [4 *šá* GABA-*šú* KUR]
6'   *ina* ITU.GAN UD.15.KA[M ᵈUTU *ina* UGU 10 UŠ *ár* MUL.*kin- ṣi* ŠÚ-*ma*]
7'   ⌜*ina* ½⌝ DANNA EGIR MU[L.*e₄-ru₆* KUR]

8'   *ina* ITU.AB UD.15.KAM ^{d}UTU [*ina* UGU 4 *šá* MUL.LU.LIM ŠÚ-*ma*]
9'   *ina* UGU MUL *kip-p*[*at* KUR]
10'   [*ina* I]TU.ZÍZ UD.15.KAM [^{d}UTU *ina* UGU MUL.*na-aṣ-ra-pi* ŠÚ-*ma*]
11'   [*ina*] ⌜UGU⌝ [MUL *šá tak-šá-a-tum* KUR]

Critical Apparatus and Philological Notes

1'   The sign UGU is omitted after *ina*.
3'   The sign UGU is omitted after *ina*.
5'   The sign UGU is omitted after *ina*.
7'   The sign UGU is omitted after *ina*. In place of *ár* which is found elsewhere in
      the text, in this line the scribe has used EGIR. BM 34639 Obv. II 10' has the
      expected *ár*.
9'   BM 34639 Obv. II 12' writes *muḫ-ḫi* instead of UGU (uniquely in this line).

3.2.3 W 22281a

The tablet W 22281a was excavated from the so-called "house of the *āšipus*" in
Uruk. This house was occupied by two families of scholars, the Šangû-Ninurta
during the fifth and early fourth centuries BC and the Ekur-zākir family during the
late fourth and early third centuries BC.[4] Unfortunately, the tablet was found in a
disturbed context due to the digging of a later grave and so it is not possible to
assign the tablet to a specific archive.

   W 22281a was published with transliteration, German translation and copy by
Hunger (1976) as SpTU I 95. A summary of the content of the text with a brief
discussion of its underlying astronomy was given by Hunger and Pingree (1999:
99). They correctly recognized that the text contains lists of constellations which
cross the meridian (i.e., culminate) at sunrise on the 15th day of each month in the
schematic calendar,[5] but did not link the text with the rising time schemes.

   The following transliteration is based upon Hunger's edition. I have restored
some entries based upon the full reconstructed rising time scheme.

---

[4]For details of the archives found at this house, see Clancier (2009). The tablets have been
published by H. Hunger and E. von Weiher in the five-volume *Spätbabylonische Texte aus Uruk*
(SpTU) series. For a study of the astronomical tablets found in the house, see Steele (forthcoming).
[5]Hunger and Pingree (1999: 99) write that the stars cross the meridian "just before sunrise",
presumably because the stars would not be visible at sunrise itself because the sky would be too
bright. However, the list is not a record of observations but rather part of a mathematically derived
scheme and so invisibility of the stars at sunrise is not significant.

Transliteration

1' DIŠ *ina* ITU.BÁR UD.1[5.KAM KI KUR *šá* ^dUTU MUL.*ku-mar šá*
MUL UD.KA.DUḪ.A]

2' MURUB₄ MUL.SUḪUR.[MAŠ *šit-qu-lu ina* ITU.DU₆ UD.15.KAM *ina li-la-
a-ti* KI ŠÚ *šá* ^dUTU ŠU.BI.AŠ.ÀM]

- -

3' DIŠ *ina* ITU.GU₄ UD.15.KAM KI KUR [*šá* ^dUTU 10 UŠ *ár* MUL.SA4 *šá*
GABA-*šú* ...]

4' *u* MUL.GU.LA *šit-qu-lu ina* IT[U.A]PIN <UD.15.KAM> [*ina*] *l*[*i-l*]*a-a-tú* KI
ŠÚ *šá* ^d[UTU ŠU.BI.AŠ.ÀM]

- -

5' DIŠ *ina* ITU.SIG₄ UD.15.KAM KI KUR *šá* ^dUTU *še-pe-e-ti* MUL.UD.KA.
DUḪ.A MUL.IKU *u*

6' MUL.ŠIM.MAḪ *šit-qu-lu ina* ITU.GAN <UD.15.KAM> *ina li-la-a-ti* KI ŠÚ
šá ^dUTU ŠU.BI.AŠ.À[M]

- -

7' DIŠ *ina* ITU.ŠU UD.15.KAM KI KUR *šá* ^dUTU SI MUL.*lu-lim* MUL.*A-ni-
ni-tum*

8' *u* MURUB₄ MUL.KU₆ *šit-qu-lu ina* ITU.AB UD.15.KAM *ina li-la-a-ti* KI
ŠÚ *šá* ^dUTU ŠU.BI.AŠ.ÀM

- -

9' DIŠ *ina* ITU.IZI UD.15.KAM KI KUR *šá* ^dUTU MUL.*na-aṣ-ra-*<*pu*> MUL.
ku-ma-ri MUL.ŠE.G[I]

10' ^dBIL.GI ZALÁG *šá* IGI ^d*en-me-šár-ra u* MURUB₄ MUL.LÚ.ḪUN.GÁ *šit-
qu-lu*

11' *ina* ITU.ZÍZ UD.15.KAM *ina li-la-a-ti* KI ŠÚ *šá* ^dUTU ŠU.BI.AŠ.ÀM

- -

12' DIŠ *ina* ITU.KIN UD.15.KAM KI KUR *šá* ^dUTU ½ DANNA EGIR MUL.
KIŠIB GÀM.A *ziq-pi* GUB[?]-*az-ma* MU[L ...]

13' MUL.MAŠ.TAB.BA MUL.SIPA.ZI.AN.NA *šit-qu-lu ina* ITU.ŠE UD.15.K
[AM]

14' *ina li-la-a-ti* KI ŠÚ *šá* ^dUTU ŠU.BI.AŠ.À[M]

- -

15' [DIŠ *ina*] I[TU.DU₆ UD.1]5.KAM KI KUR *šá* ^dUTU 5 UŠ EGIR MUL.A[L.
L]UL x *ziq-pi* G[UB-*az-ma* ...]

16' [... MU]L.[KA]K.SI.S[Á x] x [...]

Translation

1'   ¶ Month I, day 1[5 with the rising of the Sun, The Shoulder of the Panther]
2'   the middle of The Goat-[fish are in balance ...]

- -

3'   ¶ Month II, day 15, with the rising [of the Sun, 10 UŠ behind the Bright Star of
      its Chest ...]
4'   and The Great One are in balance. Mon[th V]III, <day 15>, [in] the ev[en]ing
      with the setting of the [Sun it is the same.]

- -

5'   ¶ Month III, day 15, with the rising of the Sun, The Foot of The Panther, The
      Field and
6'   The Swallow are in balance. Month IX, <day 15>, in the evening with the
      setting of the Sun it is the same.

- -

7'   ¶ Month IV, day 15, with the rising of the Sun, The Horn of the Stag,
      Anunitum
8'   and the middle of The Fish are in balance. Month X, day 15, in the evening
      with the setting of the Sun it is the same.

- -

9'   ¶ Month V, day 15, with the rising of the Sun, *Naṣrapu*, The Shoulder of the
      Old Ma[n,]
10'   Girru, the Bright (star) which is in Front of Enmešarra and the middle of the
      Hired Man are in balance.
11'   Month XI, day 15, in the evening with the setting of the Sun it is the same.

- -

12'   ¶ Month VI, day 15, with the rising of the Sun, ½ *bēru* behind The Handle of
      the Crook stands at its culmination and [...]
13'   The Twins, The True Shepherd of Anu are in balance. Month XII, day 15,
14'   in the evening with the setting of the Sun it is the sam[e.]

- -

15'   [¶] Month VII, day 1]5, with the rising of the Sun, 5 UŠ behind The Cr[a]b st
      [ands] at its culmination [...]
16'   [... The Ar]ro[w ...]

Critical Apparatus and Philological Notes

9'–10' It is difficult to understand how many stars are given in this entry. In MUL. APIN I i 3, The Old Man is associated with the god Enmešarra. I therefore assume that the phrase ᵈBIL.GI ZALÁG *šá* IGI ᵈ*en-me-šár-ra* refers to a separate star named for the god Girru which is described as being a "bright" star in front of the Old Man. This implies that there are four stars in this list: the *Ziqpu* Star *Naṣrapu*, The Old Man, a star named for the god Girru, and the Middle of the Hired Man.

12' Rather than the phrase *ana ziq-pi* DU "goes to its culmination" which is used in most texts to refer to culmination, the text has *ziq-pi* GUB-*az* "stands at its culmination" (note: GUB and DU are the same sign, but the phonetic complement -*az* ensures the reading GUB).

Commentary

This scheme presented on this text lists several stars which are "in balance" (*šit-qu-lu*) at the moment of the rising of the Sun. The list begins either with a *Ziqpu* Star or a distance behind a *Ziqpu* Star; only in the latter case is an explicit reference made to culmination, but it can be assumed that in the other entries the star is also culminating. The fully preserved entries for the first star in the list are summarized in the second column of Table 3.2.

The distances in UŠ between the entries in Months IV and V, Months V and VI, and Months VI and VII can be derived from the *Ziqpu* Star lists and is in each case 40 UŠ, in agreement with what we found from the simple monthly scheme texts. The star given in Month III, known as The Foot of the Panther, is not previously attested as a *Ziqpu* Star.[6] However, the Panther's Knee and Heel are common *Ziqpu* Stars. If we assume that the Foot of the Panther was situated halfway between the Knee and the Heel (which is quite possible depending upon the exact orientation of the constellation to the celestial equator), then we find a distance between the Foot of the Panther and the Horn of the Stag (the initial star given for Month IV) of 20 UŠ. As we will see, this distance is in accord with what we would expect.

Table 3.2 also shows the stars which are said to be in balance with the initial stars for each month. As recognized already by Hunger and Pingree (1999: 99), these stars are assumed to culminate at the same time as the initial star. As they noted, there are similarities between this material and the so-called GU text (BM 78161).[7] The GU text gives lists of entries, always beginning either with a *Ziqpu* Star or a point a stated distance behind a *Ziqpu* Star and followed by up to three other stars which are said to be "in a string" (GU). The entries for Months VI and VII in our text partially or wholly list the same stars as given with the equivalent *Ziqpu* Star in the GU text (the relevant entries for the other months are not

---

[6]The so-called DAL.BA.AN.NA text includes entries for the Right Foot of the Panther and the Toe of the Right Foot of the Panther. See Walker (1995) and Hunger and Pingree (1999: 105–106).

[7]The GU text is edited and analysed by Pingree and Walker (1988) and further discussed by Hunger and Pingree (1999: 90–97).

Table 3.2 Preserved entries in the extended monthly rising time scheme on W 22281a

Date	Initial (culminating) star	Stars in balance
I 15	[...]	Middle of The Goat-fish, [...]
II 15	[...]	[...], The Great One
III 15	The Foot of the Panther	The Field, The Swallow
IV 15	The Horn of the Stag	Anunitum, Middle of the Fish
V 15	*Naṣrapu*	The Shoulder of the Old Man, Girru, the Bright Star which is in Front of Enmešarra, Middle of the Hired Man
VI 15	½ *bēru* behind The Handle of the Crook	[...], The Twins, The True Shepherd of Anu
VII 15	5 UŠ behind The Crab	[...], The Arrow, [...]

preserved on the GU text). It may be particularly significant that the GU text includes an entry for 5 UŠ behind the Crab as well as one for the Crab itself (see further the discussion in Sect. 5.3).

Each section ends with a statement that the same stars culminate and are in balance at sunset 6 months later. This statement is in agreement with the data given explicitly for sunset in the two simple monthly scheme texts discussed above and is, according to the assumptions of schematic astronomy, astronomically correct.

3.2.4 Summary of Data in the Monthly Schemes

None of the three tablets containing the monthly scheme preserve the whole of the scheme. However, because entries from sunset are identical to those at sunrise six months earlier, by combining the sunrise and sunset entries from the three tablets, it is possible to reconstruct the entries for nine out of the 12 months. The result is shown in Table 3.3. Between the 15th of Month IV and the 15th of Month X, that is, between the summer and the winter solstice, the culminating point increases by 40 UŠ per month. Only three entries are preserved for the other half of the year. The three entries are all 20 UŠ, strongly suggesting that between the 15th of Month X and the 15th of Month IV, that is, between winter solstice and summer solstice, the culminating point increases by 20 UŠ per month.

3.3 The Daily Scheme

As first recognized by Schaumberger (1955), five fragments of a tablet from Uruk preserve statements of the culminating point at sunset and sunrise for each day of the year. As I will show, Schaumberger's reconstruction of the tablet is incorrect

Table 3.3 Summary of the rising time data reconstructed from the three monthly scheme texts

Date	Culminating point at sunrise	Distance to culminating point in the next month
I 15	[...]	[...]
II 15	[...]	[...]
III 15	10 UŠ behind The Knee/The Foot of the Panther	20 UŠ
IV 15	The 4 (Stars) of the Stag	40 UŠ
V 15	*Naṣrapu*	40 UŠ
VI 15	½ *bēru* behind The Handle of the Crook	40 UŠ
VII 15	5 UŠ behind The Crab	40 UŠ
VIII 15	½ *bēru* behind The 4 Stars of his Breast	40 UŠ
IX 15	½ *bēru* behind The Frond	40 UŠ
X 15	The Circle	20 UŠ
XI 15	The Star of the Triplets	20 UŠ
XII 15	[...]	[...]

and as a consequence his understanding of the scheme was wrong. The scheme is fully in agreement with the monthly scheme discussed in the previous section.

3.3.1 A 3414 (+) U 181a (+) U 181b (+) U 181c (+) U 181d

These five fragments were found at the site of the Bīt Reš temple in Uruk. Four fragments were excavated by the German team lead by J. Jordan in the 1912–1913 season and are now in the Istanbul museum. These four fragments were recognized to be from the same tablet by F.R. Kraus and numbered as U 181a–d. One fragment from the same tablet found its way via the antiquities market to Chicago along with several other tablets from the Bīt Reš and is in the collection of the Oriental Institute of the University of Chicago. Schaumberger (1955) recognized that the four fragments in Istanbul and the fragment in Chicago were part of the same tablet. Working from photographs of the Istanbul fragments sent by Kraus to O. Neuegbauer, Schaumberger attempted to reconstruct the tablet and identify its contents. He recognized that the tablet contained three columns on each side and that each column contained a series of statements for successive days in the schematic calendar. Each statement gives a position at or behind a *Ziqpu* Star which culminates at sunset and another which culminates at sunrise. Schaumberger further concluded that the culminating point at sunset always increases by 1;20 UŠ per day and that at sunrise always increases by 0;40 UŠ per day. As we shall see, this is only true in one half of the year. Schaumberger failed to recognize this point and as a result, he reconstructed the scheme in such a way that the culminating point at sunset eventually moved beyond that at sunrise, which makes no sense.

Schaumberger's error is either caused by or resulted in him misplacing the various fragments in his reconstruction.[8]

The tablet is set out in three columns, each of which contain approximately 60 lines covering a period of 2 months. Although at first sight, the tablet appears to be set out in a table, in fact the columns are presented as lists of one-line statements. Entries in neighbouring columns do not correspond to one another and in fact are not lined up. As a result, the number of lines in a column varies between 57 and 63. The tablet begins at the top of the first column of the obverse with the entry for the 15th of Month IV and then progresses through 180 days until the 14th of Month X on the obverse with the reverse containing the entries for the 15th of Month X to the 14th of Month IV. Thus the obverse covers the half year between summer and winter solstice and the reverse the half year from winter to summer solstice.

Of the five fragments, only U 181a preserves text on both sides. U 181a also preserves part of the upper edge; none of the other fragments preserve any edges. A 3414 joins the upper part of the Rev. II of U 181a. U 181c almost joins the top of A 3414 and preserves parts of Rev. II and III. U 181b almost joins the top of U 181b and preserves entries from Rev. I and II. The small fragment U 181d cannot be placed with certainty. The lines preserved on the five fragments are summarized below:

A 3414: Rev. II 28–50
U 181a: Obv. II 1–15, Rev. I 53–60, Rev. II 48–63, Rev. III 49–56
U 181b: Rev. I 3–19, Rev. II 3–16
U 181c: Rev. II 15–25, Rev. III 15–25
U 181d: Possibly Obv. I 25–33, Obv. III 40–48, Rev. II 25–33 or Rev. III 37–45 (omitted from the transliteration below)

The following transliteration attempts to mirror the layout of the tablet by leaving blank space where the repeating name of the *Ziqpu* Star has been omitted before the signs ŠÚ and KUR. In general, the signs in each row are in line with the same signs in the lines above and below. Because of the entries follow a simple scheme, it is possible to restore the whole tablet, which I have done below. Note, however, that although the identity of the *Ziqpu* Stars is certain, the writing of their names, which varies considerably between different texts,[9] may be different to how I have restored them here.

---

[8] A detailed discussion of the placement of the fragments and a full edition of the text may be found in the publication of the astronomical and related tablets in Istanbul currently being prepared by Christine Proust and myself (Steele and Proust forthcoming).
[9] See Steele (2014) for a discussion of the wide variations in the writing of the names of the *Ziqpu* Stars and its possible significance.

Transliteration

Obv. I

1	[ŠU 15 *ina muḫ-ḫi kip-pat*	ŠÚ-*ma ina muḫ-hi* 10 *ár a-si-du*	KUR]
2	[16 *ina muḫ-ḫi* 0,40 *ár*	ŠÚ-*ma ina muḫ-hi* 1,20 *ár* 4 *šá lu-lim*	KUR]
3	[17 *ina muḫ-ḫi* 1,20 *ár*	ŠÚ-*ma ina muḫ-hi* 2,40 *ár*	KUR]
4	[18 *ina muḫ-ḫi* 2 *ár*	ŠÚ-*ma ina muḫ-hi* 4 *ár*	KUR]
5	[19 *ina muḫ-ḫi* 2,40 *ár*	ŠÚ-*ma ina muḫ-hi* 5,20 *ár*	KUR]
6	[20 *ina muḫ-ḫi* 3,20 *ár*	ŠÚ-*ma ina muḫ-hi* 6,40 *ár*	KUR]
7	[21 *ina muḫ-ḫi* 4 *ár*	ŠÚ-*ma ina muḫ-hi* 8 *ár*	KUR]
8	[22 *ina muḫ-ḫi* 4,40 *ár*	ŠÚ-*ma ina muḫ-hi* 9,20 *ár*	KUR]
9	[23 *ina muḫ-ḫi* 5,20 *ár*	ŠÚ-*ma ina muḫ-hi* 10,40 *ár*	KUR]
10	[24 *ina muḫ-ḫi* 6 *ár*	ŠÚ-*ma ina muḫ-hi* 12 *ár*	KUR]
11	[25 *ina muḫ-ḫi* 6,40 *ár*	ŠÚ-*ma ina muḫ-hi* 13,20 *ár*	KUR]
12	[26 *ina muḫ-ḫi* 7,20 *ár*	ŠÚ-*ma ina muḫ-hi* 14,40 *ár*	KUR]
13	[27 *ina muḫ-ḫi* 8 *ár*	ŠÚ-*ma ina muḫ-hi* 1 *ár um-mu-lu-tú*	KUR]
14	[28 *ina muḫ-ḫi* 8,40 *ár*	ŠÚ-*ma ina muḫ-hi* 2,20 *ár*	KUR]
15	[29 *ina muḫ-ḫi* 9,20 *ár*	ŠÚ-*ma ina muḫ-hi* 3,40 *ár*	KUR]
16	[30 *ina muḫ-ḫi* 10 *ár*	ŠÚ-*ma ina muḫ-hi* 5 *ár*	KUR]
17	[IZI 1 *ina muḫ-ḫi* 10,40 *ár*	ŠÚ-*ma ina muḫ-hi* 6,20 *ár*	KUR]
18	[2 *ina muḫ-ḫi* 11,20 *ár*	ŠÚ-*ma ina muḫ-hi* 7,40 *ár*	KUR]
19	[3 *ina muḫ-ḫi* 12 *ár*	ŠÚ-*ma ina muḫ-hi* 9 *ár*	KUR]
20	[4 *ina muḫ-ḫi* 12,40 *ár*	ŠÚ-*ma ina muḫ-hi* 10,20 *ár*	KUR]
21	[5 *ina muḫ-ḫi* 13,20 *ár*	ŠÚ-*ma ina muḫ-hi* 11,40 *ár*	KUR]
22	[6 *ina muḫ-ḫi* 14 *ár* Š	Ú-*ma ina muḫ-hi* 13 *ár*	KUR]
23	[7 *ina muḫ-ḫi* 14,40 *ár*	ŠÚ-*ma ina muḫ-hi* 14,20 *ár*	KUR]
24	[8 *ina muḫ-ḫi* 0,20 *ár* MÚL *šá maš-a-ti*	ŠÚ-*ma ina muḫ-hi* 15,40 *ár*	KUR]
25	[9 *ina muḫ-ḫi* 1 *ár*	ŠÚ-*ma ina muḫ-hi* 2 *ár* SA₄ *šá* ŠU.GI	KUR]
26	[10 *ina muḫ-ḫi* 1,40 *ár*	ŠÚ-*ma ina muḫ-hi* 3,20 *ár*	KUR]
27	[11 *ina muḫ-ḫi* 2,20 *ár*	ŠÚ-*ma ina muḫ-hi* 4,40 *ár*	KUR]
28	[12 *ina muḫ-ḫi* 3 *ár*	ŠÚ-*ma ina muḫ-hi* 6 *ár*	KUR]
29	[13 *ina muḫ-ḫi* 3,40 *ár*	ŠÚ-*ma ina muḫ-hi* 7,20 *ár*	KUR]
30	[14 *ina muḫ-ḫi* 4,20 *ár*	ŠÚ-*ma ina muḫ-hi* 8,40 *ár*	KUR]
31	[15 *ina muḫ-ḫi* 5 *ár*	ŠÚ-*ma ina muḫ-hi* 10 *ár*	KUR]
32	[16 *ina muḫ-ḫi* 5,40 *ár*	ŠÚ-*ma ina muḫ-hi* 1,20 *ár na-aṣ-ra-pi*	KUR]
33	[17 *ina muḫ-ḫi* 1,20 *ár* MÚL *šá taš-ka-a-ti*	ŠÚ-*ma ina muḫ-hi* 2,40 *ár*	KUR]
34	[18 *ina muḫ-ḫi* 2 *ár*	ŠÚ-*ma ina muḫ-hi* 4 *ár*	KUR]
35	[19 *ina muḫ-ḫi* 2,40 *ár*	ŠÚ-*ma ina muḫ-hi* 5,20 *ár*	KUR]
36	[20 *ina muḫ-ḫi* 3,20 *ár*	ŠÚ-*ma ina muḫ-hi* 6,40 *ár*	KUR]
37	[21 *ina muḫ-ḫi* 4 *ár*	ŠÚ-*ma ina muḫ-hi* 8 *ár*	KUR]
38	[22 *ina muḫ-ḫi* 4,40 *ár*	ŠÚ-*ma ina muḫ-hi* 9,20 *ár*	KUR]
39	[23 *ina muḫ-ḫi* 5,20 *ár*	ŠÚ-*ma ina muḫ-hi* 10,40 *ár*	KUR]
40	[24 *ina muḫ-ḫi* 6 *ár*	ŠÚ-*ma ina muḫ-hi* 12 *ár*	KUR]
41	[25 *ina muḫ-ḫi* 6,40 *ár*	ŠÚ-*ma ina muḫ-hi* 13,20 *ár*	KUR]
42	[26 *ina muḫ-ḫi* 7,20 *ár*	ŠÚ-*ma ina muḫ-hi* 14,40 *ár*	KUR]
43	[27 *ina muḫ-ḫi* 8 *ár*	ŠÚ-*ma ina muḫ-hi* 1 *ár* MÚL GÀM	KUR]
44	[28 *ina muḫ-ḫi* 8,40 *ár*	ŠÚ-*ma ina muḫ-hi* 2,20 *ár*	KUR]
45	[29 *ina muḫ-ḫi* 9,20 *ár*	ŠÚ-*ma ina muḫ-hi* 3,40 *ár*	KUR]
46	[30 *ina muḫ-ḫi* 10 *ár*	ŠÚ-*ma ina muḫ-hi* 5 *ár*	KUR]

47	[KIN 1 *ina muḫ-ḫi* 10,40 *ár*	ŠÚ-*ma ina muḫ-hi* 6,20 *ár*	KUR]
48	[2 *ina muḫ-ḫi* 1,20 *ár* MÚL. DELE	ŠÚ-*ma ina muḫ-hi* 7,40 *ár*	KUR]
49	[3 *ina muḫ-ḫi* 2 *ár*	ŠÚ-*ma ina muḫ-hi* 9 *ár*	KUR]
50	[4 *ina muḫ-ḫi* 2,40 *ár*	ŠÚ-*ma ina muḫ-hi* 10,20 *ár*	KUR]
51	[5 *ina muḫ-ḫi* 3,20 *ár*	ŠÚ-*ma ina muḫ-hi* 1,40 *ár* KIŠIB GÀM	KUR]
52	[6 *ina muḫ-ḫi* 4 *ár*	ŠÚ-*ma ina muḫ-hi* 3 *ár*	KUR]
53	[7 *ina muḫ-ḫi* 4,40 *ár*	ŠÚ-*ma ina muḫ-hi* 4,20 *ár*	KUR]
54	[8 *ina muḫ-ḫi* 5,20 *ár*	ŠÚ-*ma ina muḫ-hi* 5,40 *ár*	KUR]
55	[9 *ina muḫ-ḫi* 6 *ár*	ŠÚ-*ma ina muḫ-hi* 7 *ár*	KUR]
56	[10 *ina muḫ-ḫi* 6,40 *ár*	ŠÚ-*ma ina muḫ-hi* 8,20 *ár*	KUR]
57	[11 *ina muḫ-ḫi* 7,20 *ár*	ŠÚ-*ma ina muḫ-hi* 9,40 *ár*	KUR]
58	[12 *ina muḫ-ḫi* 8 *ár*	ŠÚ-*ma ina muḫ-hi* 11 *ár*	KUR]
59	[13 *ina muḫ-ḫi* 8,40 *ár*	ŠÚ-*ma ina muḫ-hi* 12,20 *ár*	KUR]
60	[14 *ina muḫ-ḫi* 9,20 *ár*	ŠÚ-*ma ina muḫ-hi* 13,40 *ár*	KUR]
61	[15 *ina muḫ-ḫi* 10 *ár*	ŠÚ-*ma ina muḫ-hi* 15 *ár*	KUR]
62	[16 *ina muḫ-ḫi* 10,40 *ár*	ŠÚ-*ma ina muḫ-hi* 16,20 *ár*	KUR]

Obv. II

1	[17 *ina muḫ-ḫi* 1,20 *ár* ^dGAŠAN].TIN	ŠÚ-*ma ina* ⌜*muḫ*⌝-[*hi* 17,40 *ár*	KUR]
2	[18 *ina muḫ-ḫi* 2 *ár*]	ŠÚ-*ma ina* ⌜*muḫ*⌝-[*hi* 19 *ár*	KUR]
3	[19 *ina muḫ-ḫi* 2,40 *ár*]	ŠÚ-*ma ina* ⌜*muḫ*⌝-[*hi* 20,20 *ár*	KUR]
4	[20 *ina muḫ-ḫi* 3,20 *ár*]	⌜ŠÚ⌝-*ma ina muḫ*-[*hi* 21,40 *ár*	KUR]
5	[21 *ina muḫ-ḫi* 4 *ár*	ŠÚ]-*ma ina muḫ*-[*hi* 23 *ár*	KUR]
6	[22 *ina muḫ-ḫi* 4,40 *ár*	ŠÚ]-*ma ina muḫ*-[*hi* 24,20 *ár*	KUR]
7	[23 *ina muḫ-ḫi* 5,20 *ár*	ŠÚ-*ma ina muḫ-hi* 25,40 *ár*	KUR]
8	[24 *ina muḫ-ḫi* 6 *ár*	ŠÚ-*ma ina muḫ-hi* 2 *ár* MAŠ IGI	KUR]
9	[25 *ina muḫ-ḫi* 6,40 *ár*	ŠÚ]-⌜*ma ina muḫ*⌝-[*hi* 4,20 *ár*	KUR]
10	[26 *ina muḫ-ḫi* 7,20 *ár*	ŠÚ]-⌜*ma ina muḫ*⌝-[*hi* 5,40 *ár*	KUR]
11	[27 *ina muḫ-ḫi* 8 *ár*	ŠÚ-*ma*] ⌜*ina muḫ*⌝-[*hi* 1 *ár* MAŠ	KUR]
12	[28 *ina muḫ-ḫi* 8,40 *ár*	ŠÚ-*ma*] ⌜*ina muḫ*⌝-[*hi* 2,20 *ár*	KUR]
13	[29 *ina muḫ-ḫi* 9,20 *ár*	ŠÚ-*ma*] ⌜*ina muḫ*⌝-[*hi* 3,40 *ár*	KUR]
14	[30 *ina muḫ-ḫi* 10 *ár*	ŠÚ-*ma ina muḫ-hi* 5 *ár*	KUR]
15	[DU₆ 1 *ina muḫ-ḫi* 10,40 *ár*	ŠÚ-*ma ina muḫ-hi* 6,20 *ár*	KUR]
16	[2 *ina muḫ-ḫi* 11,20 *ár*	ŠÚ-*ma ina muḫ-hi* 7,40 *ár*	KUR]
17	[3 *ina muḫ-ḫi* 12 *ár*	ŠÚ-*ma ina muḫ-hi* 9 *ár*	KUR]
18	[4 *ina muḫ-ḫi* 12,40 *ár*	ŠÚ-*ma ina muḫ-hi* 10,20 *ár*	KUR]
19	[5 *ina muḫ-ḫi* 13,20 *ár*	ŠÚ-*ma ina muḫ-hi* 11,40 *ár*	KUR]
20	[6 *ina muḫ-ḫi* 14 *ár*	ŠÚ-*ma ina muḫ-hi* 13 *ár*	KUR]
21	[7 *ina muḫ-ḫi* 14,40 *ár*	ŠÚ-*ma ina muḫ-hi* 14,20 *ár*	KUR]
22	[8 *ina muḫ-ḫi* 15,20 *ár*	ŠÚ-*ma ina muḫ-hi* 15,40 *ár*	KUR]
23	[9 *ina muḫ-ḫi* 16 *ár*	ŠÚ-*ma ina muḫ-hi* 17 *ár*	KUR]
24	[10 *ina muḫ-ḫi* 16,40 *ár*	ŠÚ-*ma ina muḫ-hi* 18,20 *ár*	KUR]
25	[11 *ina muḫ-ḫi* 17,20 *ár*	ŠÚ-*ma ina muḫ-hi* 19,40 *ár*	KUR]
26	[12 *ina muḫ-ḫi* 18 *ár*	ŠÚ-*ma ina muḫ-hi* 1 *ár* ALLA	KUR]
27	[13 *ina muḫ-ḫi* 18,40 *ár*	ŠÚ-*ma ina muḫ-hi* 2,20 *ár*	KUR]
28	[14 *ina muḫ-ḫi* 19,20 *ár*	ŠÚ-*ma ina muḫ-hi* 3,40 *ár*	KUR]

29	[15 *ina muḫ-ḫi* 20 *ár*	ŠÚ-*ma ina muḫ-hi* 5 *ár*	KUR]
30	[16 *ina muḫ-ḫi* 20,40 *ár*	ŠÚ-*ma ina muḫ-hi* 6,20 *ár*	KUR]
31	[17 *ina muḫ-ḫi* 1,20 *ár ku-mar*	ŠÚ-*ma ina muḫ-hi* 7,40 *ár*	KUR]
	šá MÚL.UD.KA.DUḪ.A		
32	[18 *ina muḫ-ḫi* 2 *ár*	ŠÚ-*ma ina muḫ-hi* 9 *ár*	KUR]
33	[19 *ina muḫ-ḫi* 2,40 *ár*	ŠÚ-*ma ina muḫ-hi* 10,20 *ár*	KUR]
34	[20 *ina muḫ-ḫi* 3,20 *ár*	ŠÚ-*ma ina muḫ-hi* 11,40 *ár*	KUR]
35	[21 *ina muḫ-ḫi* 4 *ár*	ŠÚ-*ma ina muḫ-hi* 13 *ár*	KUR]
36	[22 *ina muḫ-ḫi* 4,40 *ár*	ŠÚ-*ma ina muḫ-hi* 14,20 *ár*	KUR]
37	[23 *ina muḫ-ḫi* 5,20 *ár*	ŠÚ-*ma ina muḫ-hi* 15,40 *ár*	KUR]
38	[24 *ina muḫ-ḫi* 6 *ár*	ŠÚ-*ma ina muḫ-hi* 17 *ár*	KUR]
39	[25 *ina muḫ-ḫi* 6,40 *ár*	ŠÚ-*ma ina muḫ-hi* 18,20 *ár*	KUR]
40	[26 *ina muḫ-ḫi* 7,20 *ár*	ŠÚ-*ma ina muḫ-hi* 19,40 *ár*	KUR]
41	[27 *ina muḫ-ḫi* 8 *ár*	ŠÚ-*ma ina muḫ-hi* 1 *ár* SAG MÚL.A	KUR]
42	[28 *ina muḫ-ḫi* 8,40 *ár*	ŠÚ-*ma ina muḫ-hi* 2,20 *ár*	KUR]
43	[29 *ina muḫ-ḫi* 9,20 *ár*	ŠÚ-*ma ina muḫ-hi* 3,40 *ár*	KUR]
44	[30 *ina muḫ-ḫi* 10 *ár*	ŠÚ-*ma ina muḫ-hi* 5 *ár*	KUR]
45	[APIN 1 *ina muḫ-ḫi* 0,40	ŠÚ-*ma ina muḫ-hi* 6,20 *ár*	KUR]
	ár SA₄ *šá* GABA-*šú*		
46	[2 *ina muḫ-ḫi* 1,20 *ár*	ŠÚ-*ma ina muḫ-hi* 7,40 *ár*	KUR]
47	[3 *ina muḫ-ḫi* 2 *ár*	ŠÚ-*ma ina muḫ-hi* 9 *ár*	KUR]
48	[4 *ina muḫ-ḫi* 2,40 *ár*	ŠÚ-*ma ina muḫ-hi* 10,20 *ár*	KUR]
49	[5 *ina muḫ-ḫi* 3,20 *ár*	ŠÚ-*ma ina muḫ-hi* 1,40 *ár* 4 *šá* GABA-*šú*	KUR]
50	[6 *ina muḫ-ḫi* 4 *ár*	ŠÚ-*ma ina muḫ-hi* 3 *ár*	KUR]
51	[7 *ina muḫ-ḫi* 4,40 *ár*	ŠÚ-*ma ina muḫ-hi* 4,20 *ár*	KUR]
52	[8 *ina muḫ-ḫi* 5,20 *ár*	ŠÚ-*ma ina muḫ-hi* 5,40 *ár*	KUR]
53	[9 *ina muḫ-ḫi* 6 *ár*	ŠÚ-*ma ina muḫ-hi* 7 *ár*	KUR]
54	[10 *ina muḫ-ḫi* 6,40 *ár*	ŠÚ-*ma ina muḫ-hi* 8,20 *ár*	KUR]
55	[11 *ina muḫ-ḫi* 7,20 *ár*	ŠÚ-*ma ina muḫ-hi* 9,40 *ár*	KUR]
56	[12 *ina muḫ-ḫi* 8 *ár*	ŠÚ-*ma ina muḫ-hi* 11 *ár*	KUR]
57	[13 *ina muḫ-ḫi* 8,40 *ár*	ŠÚ-*ma ina muḫ-hi* 12,20 *ár*	KUR]
58	[14 *ina muḫ-ḫi* 9,20 *ár*	ŠÚ-*ma ina muḫ-hi* 13,40 *ár*	KUR]

Obv. III

1	[15 *ina muḫ-ḫi* 10 *ár*	ŠÚ-*ma ina muḫ-hi* 15 *ár*	KUR]
2	[16 *ina muḫ-ḫi* 10,40 *ár*	ŠÚ-*ma ina muḫ-hi* 16,20 *ár*	KUR]
3	[17 *ina muḫ-ḫi* 11,20 *ár*	ŠÚ-*ma ina muḫ-hi* 17,40 *ár*	KUR]
4	[18 *ina muḫ-ḫi* 12 *ár*	ŠÚ-*ma ina muḫ-hi* 19 *ár*	KUR]
5	[19 *ina muḫ-ḫi* 12,40 *ár*	ŠÚ-*ma ina muḫ-hi* 20,20 *ár*	KUR]
6	[20 *ina muḫ-ḫi* 13,20 *ár*	ŠÚ-*ma ina muḫ-hi* 1,40 *ár* 2 *šá*	KUR]
		GIŠ.KUN-*šú*	
7	[21 *ina muḫ-ḫi* 14 *ár*	ŠÚ-*ma ina muḫ-hi* 3 *ár*	KUR]
8	[22 *ina muḫ-ḫi* 14,40 *ár*	ŠÚ-*ma ina muḫ-hi* 4,20 *ár*	KUR]
9	[23 *ina muḫ-ḫi* 15,20 *ár*	ŠÚ-*ma ina muḫ-hi* 5,40 *ár*	KUR]
10	[24 *ina muḫ-ḫi* 16 *ár*	ŠÚ-*ma ina muḫ-hi* 7 *ár*	KUR]
11	[25 *ina muḫ-ḫi* 16,40 *ár*	ŠÚ-*ma ina muḫ-hi* 8,20 *ár*	KUR]
12	[26 *ina muḫ-ḫi* 17,20 *ár*	ŠÚ-*ma ina muḫ-hi* 9,40 *ár*	KUR]

13	[27 *ina muḫ-ḫi* 18 *ár*	ŠÚ-*ma ina muḫ-hi* 1 *ár* DELE *šá* KUN-*šú*	KUR]
14	[28 *ina muḫ-ḫi* 18,40 *ár*	ŠÚ-*ma ina muḫ-hi* 2,20 *ár*	KUR]
15	[29 *ina muḫ-ḫi* 19,20 *ár*	ŠÚ-*ma ina muḫ-hi* 3,40 *ár*	KUR]
16	[30 *ina muḫ-ḫi* 20 *ár*	ŠÚ-*ma ina muḫ-hi* 5 *ár*	KUR]
17	[GAN 1 *ina muḫ-ḫi* 20,40 *ár*	ŠÚ-*ma ina muḫ-hi* 6,20 *ár*	KUR]
18	[2 *ina muḫ-ḫi* 1,20 *ár kin-ṣi*	ŠÚ-*ma ina muḫ-hi* 7,40 *ár*	KUR]
19	[3 *ina muḫ-ḫi* 2 *ár*	ŠÚ-*ma ina muḫ-hi* 9 *ár*	KUR]
20	[4 *ina muḫ-ḫi* 2,40 *ár*	ŠÚ-*ma ina muḫ-hi* 10,20 *ár*	KUR]
21	[5 *ina muḫ-ḫi* 3,20 *ár*	ŠÚ-*ma ina muḫ-hi* 1,40 *ár* e_4-*ru*$_6$	KUR]
22	[6 *ina muḫ-ḫi* 4 *ár*	ŠÚ-*ma ina muḫ-hi* 3 *ár*	KUR]
23	[7 *ina muḫ-ḫi* 4,40 *ár*	ŠÚ-*ma ina muḫ-hi* 4,20 *ár*	KUR]
24	[8 *ina muḫ-ḫi* 5,20 *ár*	ŠÚ-*ma ina muḫ-hi* 5,40 *ár*	KUR]
25	[9 *ina muḫ-ḫi* 6 *ár*	ŠÚ-*ma ina muḫ-hi* 7 *ár*	KUR]
26	[10 *ina muḫ-ḫi* 6,40 *ár*	ŠÚ-*ma ina muḫ-hi* 8,20 *ár*	KUR]
27	[11 *ina muḫ-ḫi* 7,20 *ár*	ŠÚ-*ma ina muḫ-hi* 9,40 *ár*	KUR]
28	[12 *ina muḫ-ḫi* 8 *ár*	ŠÚ-*ma ina muḫ-hi* 11 *ár*	KUR]
29	[13 *ina muḫ-ḫi* 8,40 *ár*	ŠÚ-*ma ina muḫ-hi* 12,20 *ár*	KUR]
30	[14 *ina muḫ-ḫi* 9,20 *ár*	ŠÚ-*ma ina muḫ-hi* 13,40 *ár*	KUR]
31	[15 *ina muḫ-ḫi* 10 *ár*	ŠÚ-*ma ina muḫ-hi* 15 *ár*	KUR]
32	[16 *ina muḫ-ḫi* 10,40 *ár*	ŠÚ-*ma ina muḫ-hi* 16,20 *ár*	KUR]
33	[17 *ina muḫ-ḫi* 11,20 *ár*	ŠÚ-*ma ina muḫ-hi* 17,40 *ár*	KUR]
34	[18 *ina muḫ-ḫi* 12 *ár*	ŠÚ-*ma ina muḫ-hi* 19 *ár*	KUR]
35	[19 *ina muḫ-ḫi* 12,40 *ár*	ŠÚ-*ma ina muḫ-hi* 20,20 *ár*	KUR]
36	[20 *ina muḫ-ḫi* 13,20 *ár*	ŠÚ-*ma ina muḫ-hi* 21,40 *ár*	KUR]
27	[21 *ina muḫ-ḫi* 14 *ár*	ŠÚ-*ma ina muḫ-hi* 23 *ár*	KUR]
38	[22 *ina muḫ-ḫi* 14,40 *ár*	ŠÚ-*ma ina muḫ-hi* 24,20 *ár*	KUR]
39	[23 *ina muḫ-ḫi* 15,20 *ár*	ŠÚ-*ma ina muḫ-hi* 25,40 *ár*	KUR]
40	[24 *ina muḫ-ḫi* 16 *ár*	ŠÚ-*ma ina muḫ-hi* 2 *ár na-ad-dul-lum*	KUR]
41	[25 *ina muḫ-ḫi* 16,40 *ár*	ŠÚ-*ma ina muḫ-hi* 3,20 *ár*	KUR]
42	[26 *ina muḫ-ḫi* 17,20 *ár*	ŠÚ-*ma ina muḫ-hi* 4,40 *ár*	KUR]
43	[27 *ina muḫ-ḫi* 18 *ár*	ŠÚ-*ma ina muḫ-hi* 6 *ár*	KUR]
44	[28 *ina muḫ-ḫi* 18,40 *ár*	ŠÚ-*ma ina muḫ-hi* 7,20 *ár*	KUR]
45	[29 *ina muḫ-ḫi* 19,20 *ár*	ŠÚ-*ma ina muḫ-hi* 8,40 *ár*	KUR]
46	[30 *ina muḫ-ḫi* 20 *ár*	ŠÚ-*ma ina muḫ-hi* 10 *ár*	KUR]
47	[AB 1 *ina muḫ-ḫi* 20,40 *ár*	ŠÚ-*ma ina muḫ-hi* 1,20 *ár ni-i-ri*	KUR]
48	[2 *ina muḫ-ḫi* 1,20 *ár a-si-du*	ŠÚ-*ma ina muḫ-hi* 2,40 *ár*	KUR]
49	[3 *ina muḫ-ḫi* 2 *ár*	ŠÚ-*ma ina muḫ-hi* 4 *ár*	KUR]
50	[4 *ina muḫ-ḫi* 2,40 *ár*	ŠÚ-*ma ina muḫ-hi* 5,20 *ár*	KUR]
51	[5 *ina muḫ-ḫi* 3,20 *ár*	ŠÚ-*ma ina muḫ-hi* 6,40 *ár*	KUR]
52	[6 *ina muḫ-ḫi* 4 *ár*	ŠÚ-*ma ina muḫ-hi* 8 *ár*	KUR]
53	[7 *ina muḫ-ḫi* 4,40 *ár*	ŠÚ-*ma ina muḫ-hi* 9,20	KUR]
54	[8 *ina muḫ-ḫi* 5,20 *ár*	ŠÚ-*ma ina muḫ-hi* 10,40 *ár*	KUR]
55	[9 *ina muḫ-ḫi* 6 *ár*	ŠÚ-*ma ina muḫ-hi* 2 *ár na-ad-dul-lum ár*	KUR]
56	[10 *ina muḫ-ḫi* 6,40 *ár*	ŠÚ-*ma ina muḫ-hi* 3,20 *ár*	KUR]
57	[11 *ina muḫ-ḫi* 7,20 *ár*	ŠÚ-*ma ina muḫ-hi* 4,40 *ár*	KUR]
58	[12 *ina muḫ-ḫi* 8 *ár*	ŠÚ-*ma ina muḫ-hi* 6 *ár*	KUR]
59	[13 *ina muḫ-ḫi* 8,40 *ár*	ŠÚ-*ma ina muḫ-hi* 7,20 *ár*	KUR]
60	[14 *ina muḫ-ḫi* 9,20 *ár*	ŠÚ-*ma ina muḫ-hi* 8,40 *ár*	KUR]

Rev. I

1	[15 *ina muḫ-ḫi* 10 *ár*	ŠÚ-*ma ina muḫ-hi kip-pat*	KUR]
2	[16 *ina muḫ-ḫi* 1,20 *ár* MÚL.*lu-lim*	ŠÚ-*ma ina muḫ-hi* 0,40 *ár*	KUR]
3	⌜17⌝ [*ina muḫ-ḫi* 2,40 *ár*	ŠÚ-*ma ina muḫ-hi* 1,20 *ár*	KUR]
4	18 *ina* ⌜*muḫ*⌝-[ḫi 4 *ár*	ŠÚ-*ma ina muḫ-hi* 2 *ár*	KUR]
5	19 ⌜*ina*⌝ [*muḫ-ḫi* 5,20 *ár*	ŠÚ-*ma ina muḫ-hi* 2,40 *ár*	KUR]
6	20 [*ina muḫ-ḫi* 6,40 *ár*	ŠÚ-*ma ina muḫ-hi* 3,20 *ár*	KUR]
7	21 [*ina muḫ-ḫi* 8 *ár*	ŠÚ-*ma ina muḫ-hi* 4 *ár*	KUR]
8	22 [*ina muḫ-ḫi* 9,20 *ár*	ŠÚ-*ma ina muḫ-hi* 4,40 *ár*	KUR]
9	⌜23⌝ [*ina*] ⌜*muḫ-ḫi* 10⌝,40 [*ár*	ŠÚ-*ma ina muḫ-hi* 5,20 *ár*	KUR]
10	24 *ina muḫ-ḫi* 12 *ár*	⌜ŠÚ⌝-[*ma ina muḫ-hi* 6 *ár*	KUR]
11	25 *ina muḫ-ḫi* 13,20 *ár*	⌜ŠÚ⌝-[*ma ina muḫ-hi* 6,40 *ár*	KUR]
12	26 *ina muḫ-ḫi* 14,40 *ár*	[ŠÚ-*ma ina muḫ-hi* 7,20 *ár*	KUR]
13	27 *ina muḫ-ḫi* 16 *um-mu-lu-tú* [	ŠÚ-*ma ina muḫ-hi* 8 *ár*	KUR]
14	28 *ina muḫ-ḫi* 2,20 *ár*	ŠÚ-[*ma ina muḫ-hi* 8,40 *ár*	KUR]
15	29 *ina muḫ-ḫi* 3,40 *ár*	[ŠÚ-*ma ina muḫ-hi* 9,20 *ár*	KUR]
16	30 *ina muḫ-ḫi* 5 *ár*	[ŠÚ-*ma ina muḫ-hi* 10 *ár*	KUR]
17	⌜ZÍZ⌝ 1 *ina muḫ-ḫi* ⌜6,20 *ár*⌝	[ŠÚ-*ma ina muḫ-hi* 10,40 *ár*	KUR]
18	⌜2⌝ *ina muḫ-ḫi* [7,40 *ár*	ŠÚ-*ma ina muḫ-hi* 11,20 *ár*	KUR]
19	[3] *ina muḫ-*⌜*ḫi*⌝ [9 *ár*	ŠÚ-*ma ina muḫ-hi* 12 *ár*	KUR]
20	[4 *ina muḫ-ḫi* 10,20 *ár*	ŠÚ-*ma ina muḫ-hi* 12,40 *ár*	KUR]
21	[5 *ina muḫ-ḫi* 11,40 *ár*	ŠÚ-*ma ina muḫ-hi* 13,20 *ár*	KUR]
22	[6 *ina muḫ-ḫi* 13 *ár*	ŠÚ-*ma ina muḫ-hi* 14 *ár*	KUR]
23	[7 *ina muḫ-ḫi* 14,20 *ár*	ŠÚ-*ma ina muḫ-hi* 14,40 *ár*	KUR]
24	[8 *ina muḫ-ḫi* 15,40 *ár*	ŠÚ-*ma ina muḫ-hi* 15,20 *ár*	KUR]
25	[9 *ina muḫ-ḫi* 2 *ár* SA₄ *šá* ŠU.GI	ŠÚ-*ma ina muḫ-hi* 1 *ár* MÚL *šá maš-a-ti*	KUR]
26	[10 *ina muḫ-ḫi* 3,20 *ár*	ŠÚ-*ma ina muḫ-hi* 1,40 *ár*	KUR]
27	[11 *ina muḫ-ḫi* 4,40 *ár*	ŠÚ-*ma ina muḫ-hi* 2,20 *ár*	KUR]
28	[12 *ina muḫ-ḫi* 6 *ár*	ŠÚ-*ma ina muḫ-hi* 3 *ár*	KUR]
29	[13 *ina muḫ-ḫi* 7,20 *ár*	ŠÚ-*ma ina muḫ-hi* 3,40 *ár*	KUR]
30	[14 *ina muḫ-ḫi* 8,40 *ár*	ŠÚ-*ma ina muḫ-hi* 4,20 *ár*	KUR]
31	[15 *ina muḫ-ḫi* 10 *ár*	ŠÚ-*ma ina muḫ-hi* 5 *ár*	KUR]
32	[16 *ina muḫ-ḫi* 1,20 *ár* *na-aṣ-ra-pi*	ŠÚ-*ma ina muḫ-hi* 5,40 *ár*	KUR]
33	[17 *ina muḫ-ḫi* 2,40 *ár*	ŠÚ-*ma ina muḫ-hi* 1,20 *ár* MÚL *šá taš-ka-a-ti*	KUR]
34	[18 *ina muḫ-ḫi* 4 *ár*	ŠÚ-*ma ina muḫ-hi* 2 *ár*	KUR]
35	[19 *ina muḫ-ḫi* 5,20 *ár*	ŠÚ-*ma ina muḫ-hi* 2,40 *ár*	KUR]
36	[20 *ina muḫ-ḫi* 6,40 *ár*	ŠÚ-*ma ina muḫ-hi* 3,20 *ár*	KUR]
37	[21 *ina muḫ-ḫi* 8 *ár*	ŠÚ-*ma ina muḫ-hi* 4 *ár*	KUR]
38	[22 *ina muḫ-ḫi* 9,20 *ár*	ŠÚ-*ma ina muḫ-hi* 4,40 *ár*	KUR]
39	[23 *ina muḫ-ḫi* 10,40 *ár*	ŠÚ-*ma ina muḫ-hi* 5,20 *ár*	KUR]
40	[24 *ina muḫ-ḫi* 12 *ár*	ŠÚ-*ma ina muḫ-hi* 6 *ár*	KUR]
41	[25 *ina muḫ-ḫi* 13,20 *ár*	ŠÚ-*ma ina muḫ-hi* 6,40 *ár*	KUR]
42	[26 *ina muḫ-ḫi* 14,40 *ár*	ŠÚ-*ma ina muḫ-hi* 7,20 *ár*	KUR]
43	[27 *ina muḫ-ḫi* 1 *ár* GÀM	ŠÚ-*ma ina muḫ-hi* 8 *ár*	KUR]
44	[28 *ina muḫ-ḫi* 2,20 *ár*	ŠÚ-*ma ina muḫ-hi* 8,40 *ár*	KUR]
45	[29 *ina muḫ-ḫi* 3,40 *ár*	ŠÚ-*ma ina muḫ-hi* 9,20 *ár*	KUR]
46	[30 *ina muḫ-ḫi* 5 *ár*	ŠÚ-*ma ina muḫ-hi* 10 *ár*	KUR]

47	[ŠE 1 *ina muḫ-ḫi* 6,20 *ár*	ŠÚ-*ma ina muḫ-hi* 10,40 *ár*	KUR]
48	[2 *ina muḫ-ḫi* 7,40 *ár*	ŠÚ-*ma ina muḫ-hi* 1,20 *ár* MÚL. DELE	KUR]
49	[3 *ina muḫ-ḫi* 9 *ár*	ŠÚ-*ma ina muḫ-hi* 2 *ár*	KUR]
50	[4 *ina muḫ-ḫi* 10,20 *ár*	ŠÚ-*ma ina muḫ-hi* 2,40 *ár*	KUR]
51	[5 *ina muḫ-ḫi* 1,40 *ár* KIŠIB GÀM	ŠÚ-*ma ina muḫ-hi* 3,20 *ár*	KUR]
52	[6 *ina muḫ-ḫi* 3 *ár*	ŠÚ-*ma ina muḫ-hi* 4 *ár*	KUR]
53	⌜7⌝ [*ina muḫ-ḫi* 4,20 *ár*	ŠÚ-*ma ina muḫ-hi* 4,40 *ár*	KUR]
54	8 *ina* ⌜*muḫ*⌝-[*ḫi* 5,40 *ár*	ŠÚ-*ma ina muḫ-hi* 5,20 *ár*	KUR]
55	9 *ina* ⌜*muḫ*⌝-[*ḫi* 7 *ár*	ŠÚ-*ma ina muḫ-hi* 6 *ár*	KUR]
56	10 *ina* ⌜*muḫ*⌝-[*ḫi* 8,20 *ár*	ŠÚ-*ma ina muḫ-hi* 6,40 *ár*	KUR]
57	11 *ina* ⌜*muḫ*⌝-[*ḫi* 9,40 *ár*	ŠÚ-*ma ina muḫ-hi* 7,20 *ár*	KUR]
58	12 *ina* ⌜*muḫ*⌝-[*ḫi* 11 *ár*	ŠÚ-*ma ina muḫ-hi* 8 *ár*	KUR]
59	13 *ina* ⌜*muḫ*⌝-[*ḫi* 12,20 *ár*	ŠÚ-*ma ina muḫ-hi* 8,40 *ár*	KUR]
60	14 *ina* ⌜*muḫ*⌝-[*ḫi* 13,40 *ár*	ŠÚ-*ma ina muḫ-hi* 9,20 *ár*	KUR]

Rev. II

1	[15 *ina muḫ-ḫi* 15 *ár*	ŠÚ-*ma ina muḫ-hi* 10 *ár*	KUR]
2	[16 *ina muḫ-ḫi* 16,20 *ár*	ŠÚ-*ma ina muḫ-hi* 10,40 *ár*	KUR]
3	[17 *ina muḫ-ḫi* 17,40 *ár*]	⌜ŠÚ-*ma ina muḫ*⌝-[*hi* 1,20 *ár* ^dGAŠAN. TIN	KUR]
4	[18 *ina muḫ-ḫi* 19 *ár*]	⌜ŠÚ⌝-*ma ina muḫ-hi* 2 [*ár*]	KUR
5	[19 *ina muḫ-ḫi* 20,20 *ár*]	⌜ŠÚ⌝-*ma ina muḫ-hi* 2,40 ⌜*ár*⌝	KUR
6	[20 *ina muḫ-ḫi* 21,40 *ár*]	⌜ŠÚ⌝-*ma ina muḫ-hi* 3,20 *ár* ^dGAŠAN. TIN	KUR
7	[21 *ina muḫ-ḫi* 23 *ár*]	⌜ŠÚ⌝-*ma ina muḫ-hi* 4 *ár*	KUR
8	[22 *ina muḫ-ḫi* 24,20 *ár*]	⌜ŠÚ⌝-*ma ina muḫ-hi* 4,40 *ár*	KUR
9	[23 *ina muḫ-ḫi* 25,40 *ár*]	⌜ŠÚ⌝-*ma ina muḫ-hi* 5,20 *ár*	KUR
10	[24 *ina muḫ-ḫi* 2 *ár* MAŠ IGI]	⌜ŠÚ⌝-*ma ina muḫ-hi* 6 *ár*	KUR
11	[25 *ina muḫ-ḫi* 3,20 *ár*]	ŠÚ-*ma ina muḫ-hi* 6,40 *ár*	KUR
12	[26 *ina muḫ-ḫi* 4,40 *ár*]	ŠÚ-*ma ina muḫ-hi* 7,20 *ár* ^dGAŠAN.TIN	KUR
13	[27 *ina muḫ-ḫi* 1 *ár*] MAŠ	⌜ŠÚ⌝-*ma ina muḫ-hi* 8 *ár*	KUR
14	[28 *ina muḫ-ḫi* 2,20 *ár*]	⌜ŠÚ⌝-*ma ina muḫ-hi* ⌜8,40⌝ [*ár*]	KUR
15	[29 *ina*] ⌜*muḫ-ḫi*⌝ [3,40 *ár*	ŠÚ-*ma ina muḫ-hi* 9,20 *ár*]	KUR
16	[30] *ina muḫ-ḫi* [5 *ár*	ŠÚ-*ma ina muḫ-hi* 10 *ár*]	⌜KUR⌝
17	BAR 1 *ina muḫ-ḫi* 6,⌜20⌝ [*ár*	ŠÚ-*ma ina muḫ-hi* 10,40 *ár*	KUR]
18	2 *ina muḫ-ḫi* 7,⌜40⌝ [*ár*	ŠÚ-*ma ina muḫ-hi* 11,20 *ár*	KUR]
19	3 *ina muḫ-ḫi* ⌜9⌝ [*ár*	ŠÚ-*ma ina muḫ-hi* 12 *ár*	KUR]
20	4 *ina muḫ-ḫi* [10,20 *ár*	ŠÚ-*ma ina muḫ-hi* 12,40 *ár*	KUR]
21	5 *ina muḫ*-⌜*ḫi*⌝ [11,40 *ár*	ŠÚ-*ma ina muḫ-hi* 13,20 *ár*	KUR]
22	6 *ina muḫ*-[*ḫi* 13 *ár*	ŠÚ-*ma ina muḫ-hi* 14 *ár*	KUR]
23	7 *ina* ⌜*muḫ*⌝-[*ḫi* 14,20 *ár*	ŠÚ-*ma ina muḫ-hi* 14,40 *ár*	KUR]
24	8 *ina* [*muḫ-ḫi* 15,40 *ár*	ŠÚ-*ma ina muḫ-hi* 15,20 *ár*	KUR]
25	9 [*ina muḫ-ḫi* 17 *ár*	ŠÚ-*ma ina muḫ-hi* 16 *ár*	KUR]
26	[10 *ina muḫ-ḫi* 18,20 *ár*	ŠÚ-*ma ina muḫ-hi* 16,40 *ár*	KUR]
27	[11 *ina muḫ-ḫi* 19;40 *ár*	ŠÚ-*ma ina muḫ-hi* 17,20 *ár*	KUR]
28	[12] ⌜*muḫ-ḫi* 22⌝ *ár*⌝ [ALLA	ŠÚ-*ma ina muḫ-ḫi* 18 *ár*	KUR]
29	[13] *muḫ-hi* 3,20 *ár*	ŠÚ-*ma ina muḫ*-⌜*ḫi*⌝ [18,40 *ár*	KUR]

30	[14] ⌜ina⌝ muḫ-hi 4,40 ár	ŠÚ-ma ina muḫ-ḫi 1[9,20 ár	KUR]
31	[15] ina muḫ-hi 6 ár	ŠÚ-ma ina muḫ-ḫi ⌜20⌝ [ár	KUR]
32	[1]⌜6⌝ ina muḫ-hi 7,20 ár	ŠÚ-ma ina muḫ-ḫi 20,⌜40⌝ [ár	KUR]
33	[1]7 ina muḫ-hi 8,40 ár	ŠÚ-ma ina muḫ-ḫi 1?,[20 ár ku-mar šá	KUR]
		UD.KA.DUḪ.A	
34	[1]8 ⌜ina⌝ muḫ-hi 10 ár	ŠÚ-ma ina muḫ-ḫi x [2 ár	KUR]
35	19 ina muḫ-hi 11,20 ár	ŠÚ-ma ina muḫ-ḫi [2,40 ár	KUR]
36	20 ina muḫ-hi 12,40 ár	ŠÚ-ma ina muḫ-ḫi [3,20 ár	KUR]
37	21 ina muḫ-hi 14 ár	ŠÚ-ma ina muḫ-ḫi [4 ár	KUR]
38	22 ina muḫ-hi 15,20 ár	ŠÚ-ma ina muḫ-ḫi [4,40 ár	KUR]
39	23 ina muḫ-hi 16,40 ár	ŠÚ-ma ina muḫ-ḫi [5,20 ár	KUR]
40	24 ina muḫ-hi 18 ár	ŠÚ-ma ina ⌜muḫ-ḫi⌝ [6 ár	KUR]
41	25 ina muḫ-hi 19,20 SAG?	ŠÚ-ma ⌜ina⌝ [muḫ-ḫi 6,40 ár	KUR]
	MÚL.A		
42	26 ina muḫ-hi 20,40 ár	ŠÚ-ma [ina muḫ-ḫi 7,20 ár	KUR]
43	27 ina muḫ-hi 2 ár	ŠÚ-⌜ma⌝ [ina muḫ-ḫi 8 ár	KUR]
44	28 ina muḫ-hi 3,20 ár	[ŠÚ-ma ina muḫ-ḫi 8,40 ár	KUR]
45	29 ina muḫ-hi 4,40 ár	[ŠÚ-ma ina muḫ-ḫi 9,20 ár	KUR]
46	⌜30⌝ ina muḫ-hi 6 ⌜ár⌝	[ŠÚ-ma ina muḫ-ḫi 10 ár	KUR]
47	⌜GU₄⌝ 1 ina muḫ-hi 7,20 ⌜ár⌝	[ŠÚ-ma ina muḫ-ḫi 10,40 ár	KUR]
48	2 ina muḫ-ḫi 8,40 [ár	ŠÚ]-⌜ma ina muḫ⌝-[hi 1,20 ár SA₄ šá	KUR]
		GABA-šú	
49	⌜3⌝ ina muḫ-ḫi 10 [ár]	⌜ŠÚ⌝-ma ina muḫ-⌜hi⌝ [2 ár	KUR]
50	[4] ina muḫ-ḫi 11,⌜20⌝ ár	ŠÚ-ma ina muḫ-hi [2,40 ár	KUR]
51	⌜5 ina⌝ [muḫ-ḫi] 2,40 ár	ŠÚ-ma ina muḫ-hi [3,20 ár	KUR]
52	6 ina ⌜muḫ⌝-ḫi 4 ár	ŠÚ-ma ina muḫ-hi [4] ár x?	[KUR]
53	7 ina muḫ-ḫi 5,20 ár	ŠÚ-ma ina muḫ-hi ⌜4,40⌝ ár SA₄ šá	KUR]
		⌜GABA⌝-[šú	
54	8 ina muḫ-ḫi 6,40 ár	ŠÚ-ma ina muḫ-hi ⌜5⌝,20 ár	⌜KUR⌝
55	9 ina muḫ-ḫi 8 ár	ŠÚ-ma ina muḫ-hi ⌜6⌝ ár	KUR
56	10 ina muḫ-ḫi 9,20 ár	ŠÚ-ma ina muḫ-hi 6,40 ár SA₄ šá	KUR
		GABA-šú	
57	11 ina muḫ-ḫi 10,40 ár	ŠÚ-ma ina muḫ-hi 7,20 ár	KUR
58	12 ina muḫ-ḫi 12 ár	ŠÚ-ma ina muḫ-hi 8 ár	KUR
59	13 ina muḫ-ḫi 13,20 ár	ŠÚ-ma ina muḫ-hi 8,40 ár SA₄ šá	KUR
		GABA-šú	
60	[14 ina muḫ]-⌜ḫi⌝ 14,40 ár	⌜ŠÚ-ma ina muḫ-hi⌝ 9,20 ár	KUR
61	[1]5 ina ⌜muḫ-ḫi⌝ 16 ⌜ár⌝	ŠÚ-ma ina muḫ-hi⌝ 10 ár	KUR
62	[1]6 12 ina muḫ-ḫi 17,20 ár	ŠÚ-ma ina muḫ-hi 10,⌜40⌝ ár	KUR]
63	[1]7 12 ina muḫ-ḫi 18,40 ár	ŠÚ-ma ina muḫ-hi [11,20 ár	KUR]

Rev. III

1	[18 ina muḫ-ḫi 19 ár	ŠÚ-ma ina muḫ-hi 12 ár	KUR]
2	[19 ina muḫ-ḫi 20,20 ár	ŠÚ-ma ina muḫ-hi 12,40 ár	KUR]
3	[20 ina muḫ-ḫi 1,40 ár ŠA₄ šá GIŠ.KUN-šúŠÚ-ma ina muḫ-hi 13,20 ár		KUR]
4	[21 ina muḫ-ḫi 3 ár	ŠÚ-ma ina muḫ-hi 14 ár	KUR]
5	[22 ina muḫ-ḫi 4,20 ár	ŠÚ-ma ina muḫ-hi 14,40 ár	KUR]
6	[23 ina muḫ-ḫi 5,40 ár	ŠÚ-ma ina muḫ-hi 15,20 ár	KUR]
7	[24 ina muḫ-ḫi 7 ár	ŠÚ-ma ina muḫ-hi 16 ár	KUR]

8	[25 *ina muḫ-ḫi* 8,20 *ár*	ŠÚ-*ma ina muḫ-ḫi* 16,40 *ár*	KUR]
9	[26 *ina muḫ-ḫi* 9,40 *ár*	ŠÚ-*ma ina muḫ-ḫi* 17,20 *ár*	KUR]
10	[27 *ina muḫ-ḫi* 1 *ár* DELE *šá* KUN-*šú*	ŠÚ-*ma ina muḫ-ḫi* 18 *ár*	KUR]
11	[28 *ina muḫ-ḫi* 2,20 *ár*	ŠÚ-*ma ina muḫ-ḫi* 18,40 *ár*	KUR]
12	[29 *ina muḫ-ḫi* 3,40 *ár*	ŠÚ-*ma ina muḫ-ḫi* 19,20 *ár*	KUR]
13	[30 *ina muḫ-ḫi* 5 *ár*	ŠÚ-*ma ina muḫ-ḫi* 20 *ár*	KUR]
14	[SIG 1 *ina muḫ-ḫi* 6,20 *ár*	ŠÚ-*ma ina muḫ-ḫi* 10,40 *ár*	KUR]
15	[2 *ina muḫ-ḫi* 7,40 *ár*	ŠÚ-*ma ina muḫ-ḫi* 1,20 *ár kin-ṣi*]	⌈KUR⌉
16	[3 *ina muḫ-ḫi* 9 *ár*	ŠÚ-*ma ina muḫ-ḫi* 2 *ár*]	KUR
17	[4 *ina muḫ-ḫi* 10,20 *ár*	ŠÚ-*ma ina muḫ-ḫi* 2,40 *ár*]	KUR
18	[5 *ina muḫ-ḫi* 1,40 *ár* e_4-*ru$_6$*	ŠÚ-*ma ina muḫ-ḫi* 3,20 *ár kin*]-*ṣi*	KUR
19	[6 *ina muḫ-ḫi* 3 *ár*	ŠÚ-*ma ina muḫ-ḫi* 4] ⌈*ár*⌉	KUR
20	[7 *ina muḫ-ḫi* 4,20 *ár*	ŠÚ-*ma ina muḫ-ḫi* 4,40] *ár*	KUR
21	[8 *ina muḫ-ḫi* 5,40 *ár*	ŠÚ-*ma ina muḫ-ḫi* 5],20 *ár kin-ṣi*	KUR
22	[9 *ina muḫ-ḫi* 7 *ár*	ŠÚ-*ma ina muḫ-ḫi*] ⌈6⌉ *ár*	KUR
23	[10 *ina muḫ-ḫi* 8,20 *ár*	ŠÚ-*ma ina muḫ-ḫi*] ⌈6⌉,40 *ár*	KUR
24	[11 *ina muḫ-ḫi* 19,40 *ár*	ŠÚ-*ma ina muḫ-ḫi* 7],⌈20⌉ *ár*	KUR
25	[12 *ina muḫ-ḫi* 11 *ár*	ŠÚ-*ma ina muḫ-ḫi* 8] ⌈*ár*⌉ *kin-ṣi*	⌈KUR⌉
26	[13 *ina muḫ-ḫi* 12,20 *ár*	ŠÚ-*ma ina muḫ-ḫi* 8,40 *ár*	KUR]
27	[14 *ina muḫ-ḫi* 13,40 *ár*	ŠÚ-*ma ina muḫ-ḫi* 9,20 *ár*	KUR]
28	[15 *ina muḫ-ḫi* 15 *ár*	ŠÚ-*ma ina muḫ-ḫi* 10 *ár*	KUR]
29	[16 *ina muḫ-ḫi* 16,20 *ár*	ŠÚ-*ma ina muḫ-ḫi* 10,40 *ár*	KUR]
30	[17 *ina muḫ-ḫi* 17,40 *ár*	ŠÚ-*ma ina muḫ-ḫi* 11,20 *ár*	KUR]
31	[18 *ina muḫ-ḫi* 19 *ár*	ŠÚ-*ma ina muḫ-ḫi* 12 *ár*	KUR]
32	[19 *ina muḫ-ḫi* 20,20 *ár*	ŠÚ-*ma ina muḫ-ḫi* 12,40 *ár*	KUR]
33	[20 *ina muḫ-ḫi* 21,40 *ár*	ŠÚ-*ma ina muḫ-ḫi* 13,20 *ár*	KUR]
34	[21 *ina muḫ-ḫi* 23 *ár*	ŠÚ-*ma ina muḫ-ḫi* 14 *ár*	KUR]
35	[22 *ina muḫ-ḫi* 24,20 *ár*	ŠÚ-*ma ina muḫ-ḫi* 14,40 *ár*	KUR]
36	[23 *ina muḫ-ḫi* 25,40 *ár*	ŠÚ-*ma ina muḫ-ḫi* 15,20 *ár*]	⌈KUR⌉
37	[24 *ina muḫ-ḫi* 2 *ár na-ad-dul-lum*	ŠÚ-*ma ina muḫ-ḫi* 16 *ár*]	⌈KUR⌉
38	[25 *ina muḫ-ḫi* 3,20 *ár*	ŠÚ-*ma ina muḫ-ḫi* 16,40 *ár*]	KUR
39	[26 *ina muḫ-ḫi* 4,40 *ár*	ŠÚ-*ma ina muḫ-ḫi* 17,20 *ár*]	KUR
40	[27 *ina muḫ-ḫi* 6 *ár*	ŠÚ-*ma ina muḫ-ḫi* 18 *ár*]	KUR
41	[28 *ina muḫ-ḫi* 7,20 *ár*	ŠÚ-*ma ina muḫ-ḫi* 18,40 *ár*]	⌈KUR⌉
42	[29 *ina muḫ-ḫi* 8,40 *ár*	ŠÚ-*ma ina muḫ-ḫi* 19,20 *ár*]	⌈KUR⌉
43	[30 *ina muḫ-ḫi* 10 *ár*	ŠÚ-*ma ina muḫ-ḫi* 20 *ár*	KUR]
44	[ŠU 1 *ina muḫ-ḫi* 1,20 *ár ni-i-ri*	ŠÚ-*ma ina muḫ-ḫi* 0,40 *ár a-si-du*	KUR]
45	[2 *ina muḫ-ḫi* 2,40 *ár*	ŠÚ-*ma ina muḫ-ḫi* 1,20 *ár*	KUR]
46	[3 *ina muḫ-ḫi* 4 *ár*	ŠÚ-*ma ina muḫ-ḫi* 2 *ár*	KUR]
47	[4 *ina muḫ-ḫi* 5,20 *ár*	ŠÚ-*ma ina muḫ-ḫi* 2,40 *ár*	KUR]
48	[5 *ina muḫ-ḫi* 6,40 *ár*	ŠÚ-*ma ina muḫ-ḫi* 3,20 *ár*	KUR]
49	[6 *ina muḫ-ḫi* 8 *ár*	ŠÚ-*ma ina muḫ-ḫi* 4 *ár]*	⌈KUR⌉
50	[7 *ina muḫ-ḫi* 9,20 *ár*	ŠÚ-*ma ina muḫ-ḫi* 4,40 *ár]*	KUR
51	[8 *ina muḫ-ḫi* 10,40 *ár*	ŠÚ-*ma ina muḫ-ḫi* 5,20 *ár]*	KUR
52	[9 *ina muḫ-ḫi* 2 *ár na-ad-dul-lum ár*	ŠÚ-*ma ina muḫ-ḫi* 6 *ár]*	KUR
53	[10 *ina muḫ-ḫi* 3,20 *ár*	ŠÚ-*ma ina muḫ-ḫi* 6,40 *ár]*	KUR
54	[11 *ina muḫ-ḫi* 4,40 *ár*	ŠÚ-*ma ina muḫ-ḫi* 7,20 *ár a-s*]*i-du*	KUR
55	[12 *ina muḫ-ḫi* 6 *ár*	ŠÚ-*ma ina muḫ-hi* 8 *ár]*	KUR
56	[13 *ina muḫ-ḫi* 7,20 *ár*	ŠÚ-*ma ina muḫ-hi* 8,40 *ár]*	⌈KUR⌉
57	[14 *ina muḫ-ḫi* 8,40 *ár*	ŠÚ-*ma ina muḫ-hi* 9,20 *ár*	KUR]

Translation

The repetitive nature of the text makes a complete translation unwarranted. I therefore translate only the first three lines. The remainder of the text follows the same pattern.

1 [Month IV, 15, at the culmination of The Circle (the Sun) sets and at the culmination of 10 (UŠ) after The Heel (the Sun) rises.]
2 [16, at the culmination of 0;40 (UŠ) after (ditto) (the Sun) sets and at the culmination of 1;20 (UŠ) after The 4 Stars of the Stag (the Sun) rises.]
3 [17, at the culmination of 1;20 (UŠ) after (ditto) (the Sun) sets and at the culmination of 2;40 (UŠ) after (ditto) (the Sun) rises]

   etc.

Critical Apparatus and Philological Notes

Rev. I 17   Schaumberger omitted the sign ZÍZ ("Month XI") from his transcription.
Rev. II 47   Schaumberger omitted the sign GU$_4$ ("Month II") from his transcription.
Rev. III 53 Schaumberger read the star name as *e-du* instead of *a-si-du*. The DU sign is clear, but his E is a misreading of the end of SI.

Commentary

The tablet presents a scheme which gives the point behind a *Ziqpu* Star which culminates at sunset and sunrise on each day of the schematic 360-day year. The scheme as presented almost certainly begins with the 15th of Month IV, the date of the summer solstice in schematic astronomy, and each side of the tablet contains entries for 6 months spread over three columns. Thus the obverse of the tablet covered the period from the summer solstice to the winter solstice and the reverse the period from the winter solstice to the summer solstice. Only traces of a few entries of the obverse are preserved but the mirroring of entries for sunrise and sunset 6 months apart allows the whole of the scheme to be reconstructed.

The scribe usually omits the name of the *Ziqpu* Star and instead leaves a blank space as an implied "ditto" to the star name given in an earlier line. Usually, but not always, he writes the name of a star only the first time it appears. When the scheme moves past the position of a new *Ziqpu* Star the scribe usually continues giving the position relative to the old star until the position is at least 1 UŠ behind the new star.[10]

According to the scheme for the period from the 15th of Month X to the 14 of Month IV, the culminating point at sunset increases by 1;20 UŠ per day and that of sunrise increases by 0;40 UŠ per day. However, the scheme as written contains a

---

[10]These points were already recognized by Schaumberger (1955).

Table 3.4 The complete reconstructed monthly scheme

Date	Culminating point at sunset	Distance to culminating point in the next month	Culminating point at sunrise	Distance to culminating point in the next month
I 15	[5 UŠ behind The Crab]	40 UŠ	[The Shoulder of the Panther]	20 UŠ
II 15	[½ *bēru* behind The 4 Stars of his Breast]	40 UŠ	[10 UŠ behind The Bright Star of its Breast]	20 UŠ
III 15	[½ *bēru* behind The Frond]	40 UŠ	10 UŠ behind The Knee/The Foot of the Panther	20 UŠ
IV 15	[The Circle]	20 UŠ	The 4 (Stars) of the Horn of the Stag	40 UŠ
V 15	[The Star of the Triplets]	20 UŠ	*Naṣrapu*	40 UŠ
VI 15	[The Lady of Life]	20 UŠ	½ *bēru* behind The Handle of the Crook	40 UŠ
VII 15	[The Shoulder of the Panther]	20 UŠ	5 UŠ behind The Crab	40 UŠ
VIII 15	[10 UŠ behind The Bright Star of its Breast]	20 UŠ	½ *bēru* behind The 4 Stars of his Breast	40 UŠ
IX 15	10 UŠ behind [The Knee]	20 UŠ	½ *bēru* behind The Frond	40 UŠ
X 15	The 4 (Stars) [of the Horn of the Stag]	40 UŠ	The Circle	20 UŠ
XI 15	[*Naṣrapu*]	40 UŠ	The Star [of the Triplets]	20 UŠ
XII 15	[½ *bēru* behind The Handle of the Crook]	40 UŠ	[The Lady of Life]	20 UŠ

simple computational error in the lower part of Rev. II: sometime between lines 19 and 28 the scribe has increased the position by 2;20 UŠ instead of 1;20 UŠ. This error carries through to all subsequent entries for sunset in Rev. II. It is uncertain whether the scribe corrected this error at the top of Rev. III. In my restoration I have assumed that he did notice the error at the beginning of Rev. III and calculated the subsequent values correctly. The values for sunrise are calculated correctly.

Daily increases in the culminating positon for sunset and sunrise of 1;20 UŠ and 0;40 UŠ are equivalent to increases of 40 UŠ and 20 UŠ per month, exactly in accord with the monthly scheme discussed above. Furthermore, the culminating positions for sunset and sunrise on the 15th of each month are identical with the positions given in the monthly scheme, except for the scribal error mentioned above.

3.4   Conclusion

The monthly and daily schemes discussed in this chapter provide evidence for a single calendar-based rising time scheme which assign positions at or behind *Ziqpu* Stars which culminate at sunset and sunrise. In this scheme the distance between the point which culminates at sunset in successive months is equal to 40 UŠ between the middle of Month X and the middle of Month IV (i.e. between the winter solstice and the summer solstice) and is equal to 20 UŠ between the middle of Month IV and the middle of Month X (i.e. between summer solstice and winter solstice). The culminating point at sunrise is equal to the culminating point at sunset 6 months earlier. Thus, the difference between the culminating points at sunrise in two successive months is equal to 20 UŠ between winter solstice and summer solstice and 40 UŠ between summer solstice and winter solstice. Linking the scheme directly to the dates of the solstices in the schematic calendar following the tradition of MUL. APIN places this scheme firmly within the realm of schematic astronomy.

By combining the preserved entries from all of the texts and exploiting the relation between the sunset and sunrise data at 6-month intervals it is possible to reconstruct the complete scheme. Table 3.4 presents the complete reconstructed monthly scheme. As I will demonstrate in the next chapter, this calendar-based scheme directly parallels the more well attested zodiac-based scheme.

References

Brack-Bernsen L, Hunger H (2005–2006) On the "atypical astronomical cuneiform text E": a mean-value scheme for predicting lunar latitude. *Archiv für Orientforschung* 51:96–107

Clancier P (2009) *Les bibliothèques en Babylonie dans la deuxième moitié du Ier millenaire av. J.-C.*, Alter Orient und Altes Testament 363. Ugarit-Verlag, Münster

Hunger H (1976) Spätbabylonische Texte aus Uruk. Teil I. Gebr. Mann, Berlin

Hunger H, Pingree D (1999) Astral sciences in Mesopotamia. Brill, Leiden

Neugebauer O, Sachs A (1967) Some atypical astronomical cuneiform texts. I. J Cuneiform Stud 21:183–218

Pingree D, Walker C (1988) A Babylonian star catalogue: BM 78161. In: Leichty E et al (eds) A scientific humanist: studies in memory of Abraham Sachs. University Museum, Philadelphia, pp 313–322

Schaumberger J (1955) Anaphora und Aufgangskalender in neuen Ziqpu-Texten. Zietschrift für Assyriologie 52:237–251

Steele JM (2007) Celestial measurement in Babylonian astronomy. Ann Sci 64:293–325

Steele JM (2012) Remarks on the sources for the lunar latitude section of atypical astronomical cuneiform Text E, *NABU* 2012/3, no. 54, pp 71–72

Steele JM (2014) Late Babylonian *Ziqpu*-star lists: written or remembered traditions of knowledge? In: Bawanypeck D, Imhausen A (eds) Traditions of written knowledge in ancient Egypt and Mesopotamia, Alter Orient und Altes Testament 403. Ugarit-Verlag, Münster, pp 123–151

Steele JM (forthcoming) Astronomical Activity in the 'House of the *āšipus*' in Uruk. In: Proust C, Steele JM (eds) Scholars and scholarship in late Babylonian Uruk, in press

Steele JM, Proust C (forthcoming) Astronomical and related cuneiform texts from Uruk in the ancient orient museum of Istanbul, in preparation

Walker CBF (1995) The Dalbanna text: a Mesopotamian star list. Die Welt des Orients 26:27–42

Chapter 4
Zodiac-Based Rising Time Schemes

Abstract The most commonly attested form of rising time scheme is one in which culminating points at or behind *Ziqpu* Stars are given either for the beginning and end of a sign of the zodiac or for the end of each twelfth of a sign of the zodiac (known as a microzodiac sign). This chapter studies the texts containing zodiac-based rising time schemes showing that they all attest to the same basic scheme. Furthermore, it is demonstrated that this scheme is identical to the calendar-based scheme, and that the calendar-based scheme was probably the earlier of the two schemes.

Keywords Babylon · Babylonian astronomy · Culminating point · Cuneiform tablet · Uruk · *Ziqpu* stars · Zodiac

4.1 Introduction

Previous discussions of the rising time scheme have focused upon a group of texts containing what I term the microzodiac rising time scheme. This scheme assigns a culminating position at or behind a *Ziqpu* Star to each twelfth of a zodiacal sign. This division of a zodiacal sign into twelve parts is known in modern scholarship as the "microzodiac". Each microzodiac sign extends over 2½° and can be named by either (or both) a number representing its placement within the zodiacal sign using the term ḪA.LA "portion" (e.g. 3-*tú* ḪA.LA *šá* MÚL.MAŠ "3rd portion of Gemini") or by naming the twelve microzodiac signs after the twelve zodiacal signs, beginning with the same sign as the zodiacal sign that is being divided (e.g. MÚL.MAŠ *šá* MÚL.MAŠ "Gemini of Gemini" is the first portion of Gemini, MÚL.ALLA *šá* MÚL.MAŠ "Cancer of Gemini" is the second portion of Gemini, and so on up to MÚL.MÚL *šá* MÚL.MAŠ "Taurus of Gemini" which is the twelfth and final portion of Gemini". Following Monroe (2016) I designate the sign names used in the microzodiac as the major and minor signs (thus in the example "Cancer of Gemini", Gemini is the major sign and Cancer is the minor sign).

© The Author(s) 2017
J.M. Steele, *Rising Time Schemes in Babylonian Astronomy*, SpringerBriefs
in History of Science and Technology, DOI 10.1007/978-3-319-55221-7_4

The culminating point assigned to a microzodiac sign refers to the moment when the end of the 2½° microzodiacal sign rises across the eastern horizon. Thus, the culminating point given for the first microzodiacal sign within a sign corresponds to when the point 2½° within that sign is rising. The culminating point given for the 12th microzodiacal sign corresponds to the moment when the end of a zodiacal sign (i.e. 30° within the sign) and the beginning of the next sign (i.e. 0° within the next sign), or the boundary between the two signs, rises.

In addition to giving the culminating points for the microzodiacal signs, the detailed microzodiac texts also give for each entry a statement of a month, one of the three watches of the day which is equated with either the 28th, the 29th or the 30th, and a star from the repertoire of the Three Stars Each ("Astrolabe") texts which is said to "expel a flare" (for an discussion of the meaning of this term, see the philological notes to BM 34713 Rev. I 15 in Sect. 4.2.1). Each section of these texts, which concern one sign of the zodiac, also often gives a summary of the culminating points for the beginning and end of the zodiacal sign and the distance in UŠ and NINDA which culminates over the course of one microzodiac portion of 2½° and the complete 30° of the zodiacal sign. Texts which contain the section for Aries also include a long preliminary statement after the division of daylight into three sections. The Aries section starts with the middle of Aries rather than the beginning of the zodiacal sign as is found in sections for the other signs of the zodiac. This choice to begin the scheme in the middle of Aries reflects the placement of the vernal equinox in the middle of Month I (and hence at 15° of Aries) in the schematic calendar.

Several texts present a simplified version of the microzodiac scheme in which the rising time scheme is presented by itself, without the additional material concerning the months, parts of day, days of the month, and flaring stars.

Both the full microzodiac texts and the simplified texts usually include a summary of the culminating points at the beginning and end of a sign of the zodiac. One tablet contains a collection of these statements for the twelve signs of the zodiac, although entries for only the first five signs are preserved.

In Sects. 4.2–4.4 I edit the texts in these various groups before analyzing the microzodiac scheme as a whole in Sect. 4.5.

4.2   Texts Presenting the Microzodiac Scheme

Five texts are known to me to contain copies of the detailed microzodiac rising-time scheme.[1] BM 34713 and BM 34664 from Babylon and A 3427 from Uruk are all substantial fragments which collectively preserve all or part of the sections

---

[1]In addition, BM 41679 (=LBAT 1509) may contain a text of this type or may just quote from from the scheme. Unfortunately, the surface of the tablet is extremely badly damaged and I have not succeeded in making a coherent reading of more than a few lines text.

concerning Aries, Taurus, Libra, Scorpio, Sagittarius, Aquarius and Pisces. These three texts are edited and discussed here. Two other tablets BM 32276 and U 195 contain copies of the Aries section known from BM 34713. BM 32276 also contains material that is related to the rising-time schemes and is edited along with other related texts in Chap. 5. U 195 will be edited and discussed by Christine Proust and myself in our forthcoming publication of the astronomical and related cuneiform tablets from Uruk in Istanbul (Steele and Proust forthcoming). Aside from allowing a few damaged signs to be restored in BM 34713, U 195 does not add anything significant to the understanding of the rising time scheme. The existence of these duplicates of the Aries section, and the very stable format and terminology found in all five texts, whether they be from Babylon or Uruk, suggests that there existed a standard composition on the microzodiac rising time scheme of which we have parts from several copies.

4.2.1 BM 34713 Rev. I 10–34

BM 34713 (Sp. II 202 + 81–6–24, 140 + SH. 81–7–6, 703 + 82–7–4, 87), almost certainly from Babylon, is a large fragment of a two column tablet (Fig. 4.1). A copy of the tablet by Pinches is published as LBAT 1499. The fragment preserves almost exactly the right-hand half of the tablet; Obv. II and Rev. I are more or less completely preserved, but only a few traces of Rev. II remain and Obv. I is completely lost.

BM 34713 is a compilation text. The obverse and the beginning of the reverse contain a copy of a Three Star Each text.[2] Following a horizontal ruling, a few cm of blank space, and a further horizontal ruling is the beginning of a rising time text, which presumably continued into the now lost Rev. II. The rising time text has previously been discussed and edited by Schaumberger (1955) and Rochberg (2004: Text B). It is duplicated by two other sources, BM 32276 Obv. from Babylon and U 195 from Uruk, not known to these authors and which allow some small improvements to the edition.

Transliteration

10 ⌜ki⌝-i ina ITU.BÁR ITI AN.KU$_{10}$ ⌜dr⌝UTU⌝ ina UGU MÚL.[ku-mar šá] MÚL.UD.KA.DUḪ.A KUR-ḫa

11 ina muḫ-ḫi MÚL.ME ár-tú šá MÚL.ALLA ŠÚ-ma 6 KASKAL u$_4$-mu ana 3 ⌜ḪA.LA.MEŠ⌝ KAS$_5$⌝ za-az-ma

12 ḪA.LA reš-tú ⌜2⌝ DANNA še-rim UD.20 + GAR(error for: 28).KAM 2-tú ḪA.LA 2 DANNA [AN.NE] ⌜UD.29.KAM⌝

---

[2]For an edition and extensive discussion of this part of the tablet, see Horowitz (2014: 124–139).

Fig. 4.1 BM 34713 reverse

13   3-*tú* ḪA.LA 2 DANNA EN.USAN UD.30.KAM *ki-i ina* ITU.BÁR
        KI KUR *šá* ^dUTU MÚL.*ku-mar šá* MÚL.UD.KA.DUḪ

14   *ana ziq-pi* DU-*ma* ^dUTU *ki-i* GIŠ.KUN MÚL.LU AN.KU$_{10}$ TAB-*ú* 6-*tú*
        ḪA.LA *šá* MÚL.LU MÚL.ABSIN

15   *šá* MÚL.LU KIN KUR *ina* KIN *ina še-rim* UD.28.KAM MÚL.BIR *meš-ḫu*
        im-šuḫ ZI 1 UŠ 40 NINDA

16   *ár* MÚL.*ku-mar šá* MÚL.UD.KA.DUḪ *ana ziq-pi* DU-*ma šamaš* KI.MIN
        7-*tú* ḪA.LA *šá* MÚL.LU

17   MÚL.RÍN *šá* MÚL.LU DU$_6$ KUR *ina* DU$_6$ *ina še-rim* UD.28 MÚL.NIN.
        MAḪ *meš-ḫu im-šuḫ* ZI

18   3 UŠ 20 NINDA *ár* SI(error for: *ku*)-*mar* MÚL.UD.KA.DUḪ *ana ziq-pi*
        DU-*ma šamaš* KI.MIN 8-*tú* ḪA.LA

19   *šá* MÚL.LU MÚL.GÍR.TAB *šá* MÚL.LU APIN KUR *ina* APIN *ina še-rim*
      28 MÚL.UR.IDIM *meš-ḫu im-šuḫ* ZI

20   5 UŠ *ár ku-mar* MÚL.UD.KA.DUḪ *ana* ⸢*ziq*⸣-*pi* DU-*ma šamaš* KI.MIN 9-
      tú ḪA.LA *šá* MÚL.LU MÚL.PA.BIL

21   *šá* MÚL.LU GAN KUR *ina* GAN *ina še-rim* 2[8 MÚL.*ṣal*]-*bat-a-nu meš-*
      ḫu im-šuḫ ZI 6 UŠ 40 NINDA

22   *ár* MÚL.*ku-mar* MÚL.UD.KA.D[UḪ] ⸢*ana ziq-pi* DU⸣-*ma šamaš* KI.MIN
      10-*tú* ḪA.LA *šá* MÚL.LU MÚL.MÁŠ *šá* MÚL.LU

23   AB KUR *ina* AB *ina še-rim* 28 MÚL.GU.LA *meš-ḫu im-šuḫ* ZI 8 UŠ 20
      NINDA *ár* MÚL.*ku-mar*

24   MÚL.UD.KA.DUḪ *ana ziq-pi* DU-*ma šamaš* KI.MIN 11-*tú* ḪA.LA *šá*
      MÚL.LU MÚL.GU.LA *šá* MÚL.LU

25   ZÍZ KUR *ina* ZÍZ *ina še-rim* 28 MÚL.*nu-muš-da meš-ḫu im-šuḫ* ZI MÚL.
      SA₄ *šá* GABA-*šú ana* ⸢*ziq*⸣-*pi*

26   DU-*ma šamaš* KI.MIN 12-*tú* ḪA.LA *šá* MÚL.LU MÚL.AŠ.IKU *šá* MÚL.
      LU ŠE KUR *ina* ŠE *ina še-rim* 28 MÚL.KU₆

27   *meš-ḫu im-šuḫ* ZI PAP 10 UŠ TA MÚL.*ku-mar šá* MÚL.UD.KA.DUḪ EN
      SA₄ *šá* [G]ABA-*šú* MÚL LU

28   [TA GIŠ].KUN-*šú* EN TIL KUR 1-*et* ḪA.LA 1 UŠ 40 NINDA *ziq-pi i-lak-*
      ma 2 ⸢UŠ⸣ [30] ⸢NINDA⸣ 1-*et* ḪA.LA

29   [*šá* MÚL.L]U KUR *ina* 6 ḪA.LA.MEŠ 10 UŠ *ziq-pi i-lak-ma*

30   [6 ḪA].LA *šá* MÚL.LU TA MAŠ-*šú* [E]N TIL-*šú* KUR

- -

31   [TA SA₄ *šá*] GABA EN 4 UŠ *ina* IGI MÚL.*kin-si* MÚL.MÚL EN TIL⸢?⸣
      ⸢KUR⸣ 1 UŠ 40 NINDA *ár* SA₄

32   [*šá* GABA-*šú ana ziq-p*]i DU-*ma šamaš* KI.MIN ḪA.LA *reš-tú šá* MÚL.
      [MÚL] ⸢MÚL⸣.MÚL *šá* MÚL.MÚL GU₄ KUR

33   [*ina* GU₄ *ina še-rim* 28 MÚL.MÚL m]*eš-ḫa im-šuḫ* ZI 3 UŠ ⸢20⸣ [NINDA
      ár SA₄ *šá* GABA] ⸢*ana*⸣ *ziq-pi* DU-*ma*

34   [*šamaš* KI.MIN 2-*tú* ḪA.LA *š*]*á* MÚL.MÚL MÚL.MAŠ.MAŠ *šá* MÚ[L.
      MÚL SIG KUR *ina* SIG *ina še-rim* 28 MÚL.SIPA.Z]I.AN.NA

Translation

10   When in Month I, the month of the eclipse, the Sun rises at the culmination
      of The [Shoulder] of The Panther (and)

11   sets at the culmination of The Rear Stars of the Crab, and 6 *bēru* is the day
      (which) into 3 portions … is divided:

12   The first portion of 2 *bēru* is the morning (and corresponds to) the <28> th
      day. The 2nd portion of 2 *bēru* is [the noontime] (and corresponds to) the
      29th day.

13   The 3rd portion of 2 *bēru* is the afternoon (and corresponds to) the 30th
      day. When in Month I with the rising of the Sun the Shoulder of the Panther

14   culminates and the Sun, when it is at the rump of Aries begins an eclipse, (at) the 6th portion of Aries (which) is Virgo

15   of Aries (and corresponds to) Month VI rises. In Month VI, in the morning (which corresponds to) the 28th day, The Kidney expels a flare. 1 UŠ 40 NINDA

16   after The Shoulder of the Panther culminates and the Sun ditto (at) the 7th portion of Aries

17   (which) is Libra of Aries (and corresponds to) Month VII rises. In Month VII, in the morning (which corresponds to) the 28th day *Ninmaḫ* expels a flare.

18   3 UŠ 20 NINDA after The Shoulder of the Panther culminates and the Sun ditto (at) the 8th portion

19   of Aries (which) is Scorpio of Aries (and corresponds to) Month VIII rises. In Month VIII, in the morning (which corresponds to) the 28(th day), The Wolf expels a flare.

20   5 UŠ after The Shoulder of the Panther culminates and the Sun ditto (at) the 9th portion of Aries (which) is Sagittarius

21   of Aries (and corresponds to) Month IX rises. In Month IX, in the morning (which corresponds to) the 2[8(th day) M]ars expels a flare. 6 UŠ 40 NINDA

22   behind The Shoulder of the Panther culminates and the Sun ditto (at) the 10th portion of Aries (which) is Capricorn of Aries

23   (and corresponds to) Month X rises. In Month X, in the morning (which corresponds to) the 28(th day), The Great One expels a flare. 8 UŠ 20 NINDA behind The Shoulder of the

24   Panther culminates and the Sun ditto (at) the 11th portion of Aries (which) is Aquarius of Aries

25   (and corresponds to) Month XI rises. In Month XI, in the morning (which corresponds to) the 28(th day), Numušda expels a flare. The Bright Star of its Breast culminates

26   and the Sun ditto (at) the 12 portion of Aries which is Pisces of Aries (and corresponds to) Month XII rises. In Month XII, in the morning (which corresponds to) the 28(th day) The Fish

27   expels a flare. A total of 10 UŠ. From The Shoulder of the Panther to the Bright Star of its Breast, Aries

28   [from its ru]mp to (its) end rises. (For) 1 portion, 1 UŠ 40 NINDA culminates (and) 2 UŠ [30] NINDA (which is) 1 portion.

29   [of Ar]ies rises. In 6 portions, 10 UŠ culminates (and)

30   [6 port]ions of Aries from its middle [t]o its end rise.
- -
31   [From the Bright Star of] its Breast to 4 UŠ in front of The Knee, Taurus to its end rises. 1 UŠ 40 NINDA behind the Bright Star

32   [of its Breast cul]minates and the Sun ditto (at) the first portion of Tau[rus] (which is) Taurus of Taurus (and corresponds to) Month II rises.

33   [In Month II, in the morning (which corresponds to) the 28(th day) The
        Stars] expel a [f]lare. 3 UŠ 20 [NINDA behind the Bright Star of its Breast]
        culminates and

34   [the Sun ditto at the 2nd portion o]f Taurus (which) is Gemini of Tau[rus
        (and corresponds to) Month III rises. In Month III, in the morning (which
        corresponds to) the 28th(th day) The True] Shepherd of Heaven

Critical Apparatus and Philological Notes

10   I do not understand the significance of the phrase ITI AN.KU$_{10}$ "Month of
        the eclipse".

11   The scribe has written *bēru* using the logogram KASKAL rather than the
        normal writing DANNA (=KASKAL-BU) used elsewhere in the text.
        Although KASKAL is an attested writing for *bēru*, it seems more likely
        that the scribe has simply miswritten a DANNA. The end of this line is
        duplicated by U 195 Obv. 2 and BM 32276 Obv. I 2 but all three copies of
        this text are damaged to some extent at this place. Rochberg read 4 instead
        of *ana* 3, but *ana* 3 is clear on the tablet and U 195 Obv. 2 has *a-na* 3,
        confirming this reading. The signs ḪA.LA are damaged but fairly clear and
        confirmed by U 195 Obv. 2 which is undamaged at this point.
        Unfortunately, the next two signs are damaged and hard to read on all three
        copies. The first sign may simply be MEŠ, indicating that ḪA.LA is plural
        ("portions") (c.f. line 29); U 195 Obv. 2 might have a damaged ME sign
        here, also used as a plural marker. The second sign appears on U 195 Obv.
        2 to be KAS$_5$, and the traces on the present tablet are consistent with such a
        reading. Pinches drew the following sign as GAR; U 195 Obv.v 2 has a
        clear ZA sign here and collation of the present tablet confirms that the sign
        here is ZA not GAR. Thus the end of this line must mean something like "6
        bēru is the day (which) into 3 portions … is divided." I suspect that the
        broken word is an indication that the division is into equal parts, but I have
        been unable to find a reading that would allow such a translation from the
        preserved traces. Note that the ḪA.LA "portions" referred to here are the
        watches of the day, not the division of the zodiacal signs into the 12
        microzodiac signs, which confusingly are referred to using the same
        terminology.

12   The scribe has made a clear error in the writing of the number 28. Instead
        he appears to have written 20 + GAR or perhaps 24.

14   I do not understand the reference to the eclipse in middle of Aries in this
        line.

15   The phrase *meš-ḫu im-šuḫ* ZI was separated into two parts by Rochberg.
        She took *meš-ḫu im-šuḫ* "flares" to be the end of a sentence and ZI "dis-
        tance" to be the beginning of the next sentence. This reading is initially
        attractive because the phrase *meš-ḫu im-šuḫ* is used frequently in astro-
        logical omens (including omens found on the Three Stars Each text copied
        on the same tablet as the present text) to refer to a star producing some form

of glow,[3] and ZI is written before the statement of the position at or behind
a *Ziqpu* Star. Two pieces of evidence contradict this interpretation, however. First, the ZI does not appear before the first entry in either the Aries or
the Taurus sections of the text. Instead, the first entry begins just with the
name of the *Ziqpu* Star. The lack of a ZI at the beginning of the first entry
also occurs on the tablet A 3427 Obv. 2, a case where the entry begins with
a number behind the *Ziqpu* Star. The ZI sign does, however, appear at the
end of the final entry in all of these cases. More significantly, however, the
scribe of A 3427 frequently replaces the whole phrase *meš-ḫu im-šuḫ* ZI
with KI.MIN "ditto". If the phrase was to be split into the end of one
sentence and the beginning of another, I do not think that the whole thing
would be replaced by "ditto". Instead, I would expect that *meš-ḫu im-šuḫ*
would be replaced by "ditto" and the ZI written out in full. Thus, I conclude
that *meš-ḫu im-šuḫ* ZI is one phrase which therefore must refer to the
preceding star name. The term ZI = *nasāḫu* means "to tear out". Thus *meš-
ḫu im-šuḫ* ZI could mean "a glow is produced and torn out" which I
translate as "expels a flare".

16   It is unclear what *šamaš*(20) KI.MIN refers to. Earlier in the text, the Sun is
written ^dUTU and I am uncertain whether the writing 20 for the Sun here
and later is significant. KI.MIN "ditto" could perhaps refer to the rising of
the sun referred to earlier in line 13, but I interpret the KUR in line 17 as a
statement of the rising. Furthermore, it would be odd to replace KUR with
KI.MIN as the latter requires writing considerably more wedges and takes
up more space on the tablet. I suspect that it instead refers back to the
unexplained reference to an eclipse in line 14.

31   The position 4 UŠ *ina* IGI MÚL.*kin-si* "4 UŠ in front of The Knee" is
clearly written but problematical for two reasons. First, we would expect
the position to be at The Knee. Second, all other positions given on this and
other related texts give positions either at or behind (not in front of) *Ziqpu*
Stars. BM 36609 + Obv. III 18 and 20 also gives an anomalous position for
the culminating point for the end of Taurus, this time 5 UŠ in front of the
star, written in line 18 as 5 UŠ *ina* IGI *kin-ṣa* "5 UŠ in front of The Knee"
and 5 UŠ *kin-ṣa* NU IGI "5 UŠ not reaching the Knee".

Commentary

The preserved text is divided into two sections corresponding to the zodiacal signs
Aries and Taurus. Presumably, further sections for the next one or two signs of the
zodiac were given in the now lost column Rev. II. The first section begins with the
phrase "When in Month I, the month of the eclipse". In Month I the Sun is located
in the zodiacal sign Aries according to the principles of schematic astronomy. The
following statement, "the month of the eclipse", is puzzling, however. In the related
text BM 34664, which originally contained related sections for the signs Libra to

---

[3]For further discussion of this term, see Horowitz (2014: 135–139).

Pisces (corresponding to Months VII–XII), the corresponding first line reads "When in Month VII, day 15", suggesting perhaps that "the month of the eclipse" is to be understood as a reference to the 15th day, i.e., the middle of the month or one of the days on which a lunar eclipse may take place. The text continues by giving the culminating points at sunrise and sunset on that day, noting that on this day, which is the vernal equinox according to the schematic calendar, day (and therefore also night) is 6 *bēru* in length. The text then says that the day is divided into three equal parts, each of 2 *bēru*. Although the term is not used explicitly these three parts are the watches (EN.NUN) of the day. The three watches are named and then associated with one of three days: morning with the 28th, afternoon with the 29th, and evening with the 30th. These associations are used in following scheme.

The text next ties the rising of the sun in the middle of Month I to the middle of the zodiacal sign Aries which corresponds to the (end of the) 6th microzodiacal sign named Virgo of Aries. The reference to an eclipse in line 14 is again puzzling (perhaps it is meant simply to indicate that we are dealing with the middle of the month).

The microzodiac scheme proper begins at the end of line 15 and continues to line 27 with a series of statements for each microzodiac sign from the 7th to the 12th. These statements have a rigid format as shown in the following example from lines 15–17:

1 UŠ 40 NINDA *ár* MÚL.*ku-mar šá* MÚL.UD.KA.DUḪ *ana ziq-pi* DU-*ma šamaš* KI.MIN 7-*tú* ḪA.LA *šá* MÚL.LU MÚL.RÍN *šá* MÚL.LU DU₆ KUR *ina* DU₆ *ina še-rim* UD.28 MÚL.NIN.MAḪ *meš-ḫu im-šuḫ* ZI

1 UŠ 40 NINDA after The Shoulder of the Panther culminates and the Sun ditto (at) the 7th portion of Aries (which) is Libra of Aries (and corresponds to) Month VII rises. In Month VII, in the morning (which corresponds to) the 28th day *Ninmaḫ* expels a flare.

The statement begins with a position at or behind a *Ziqpu* Star which culminates when the Sun rises at a particular microzodiac sign (strictly speaking the end of the microzodiac sign).[4] The microzodiac sign is identified both by number and by name as "Minor Sign" of "Major Sign". The statement continues by equating the Minor Sign with the equivalent month. The sign KUR following the month name has been taken by Rochberg to be the beginning of a next sentence and to refer to the lunar phenomena of last visibility, which she links with the mention of days 28, 29 or 30. However, in my opinion it makes more sense to read the KUR as the verb "rises" at the end of the previous sentence and see it as referring back to the mention of the Sun before its position in the microzodiac is given. The next sentence repeats the month name derived from the Minor Sign given in the previous sentence and then gives one of the watches of the day which is associated with one of the days 28, 30 or 30 according to the rule set out above in lines 12–13. Finally, a star is said to "expel a flare", perhaps on that day. The data contained in this scheme will be discussed in Sect. 4.5.

---

[4]Rochberg (2004) incorrectly places the culminating position at the end rather than the beginning of the statement.

Following the 12th and final entry in the microzodiac scheme, lines 27–30 summarize the scheme giving the total difference in the position of the culminating point over the six microzodiac signs from the middle to the end of Aries.

The second section, which deals with Taurus, begins with a brief statement of the points which culminate at the beginning and end of Taurus. This statement reads: "[From the Bright Star of] its Breast to 4 UŠ in front of The Knee, Taurus to its end rises." This statement is problematical for two reasons. First, all other positions given on this and other related texts give positions either at or behind (not in front of) *Ziqpu* Stars. Secondly, assuming that the culminating position increases uniformly through the twelve microzodiac divisions of the sign Taurus, which is what we would expect and is what is attested in the preserved entries, the position at the twelfth and final microzodiac should be exactly at The Knee, assuming that The Knee is situated 20 UŠ behind the previous *Ziqpu* Star, The Bright Star of its Chest, as is given in all of the preserved *Ziqpu* Star lists (the 26-star list and the three 25-star lists which preserve this distance, the Sippar Planisphere, VAT 16437 and BM 38704, all give the value 20 UŠ) (Steele 2014: 142–143). BM 36609 + Obv. III 18 and 20, discussed in Sect. 4.3, also gives an anomalous position for the culminating point for the end of Taurus, this time 5 UŠ in front of the star, written in line 18 as 5 UŠ *ina* IGI *kin-ṣa* "5 UŠ in front of The Knee" and 5 UŠ *kin-ṣa* NU IGI "5 UŠ not reaching the Knee". In addition, BM 34790 (=LBAT 1502) Obv. II 17, also appears to refer to either the position 4 or 5 UŠ in front of The Knee although the context of this reference is unknown. However, BM 34639 (see Sect. 3.2.1), one of texts which presents the month-based form of the rising time scheme, gives the position "10 UŠ behind The Knee" for the 15th of Month III, which is equivalent to the 6th microzodiac portion of Gemini. This position implies that the culminating point at the end of Taurus is at (not 4 or 5 UŠ in front of) The Knee. Similarly, the related text BM 37150 (see Sect. 5.5) also gives the position 10 UŠ behind The Knee for either the middle of Month III or the middle of Taurus. I can suggest three possible explanations for this unusual entry. First, and least likely, is that it is simply a scribal error. The appearance of the same position in three different but related texts counts against such an error, however. Secondly, two versions of the rising time scheme were in circulation, one of which used a *Ziqpu* Star list in which The Knee was 24 UŠ behind The Bright Star of its Chest (which also implies that The Ankle is 6 UŠ rather than 10 UŠ behind The Knee), and the other which used the standard *Ziqpu* Star list. Thirdly, it is possible that the phrase "4/5 UŠ in front of The Knee" is in fact an alternate name for the star The Knee.[5] Counting against this explanation, however, is the two different ways of referring to the position 5 UŠ behind The Knee in BM 36609+.

---

[5]Star names which include positional reference to other stars are known in other contexts. For example, one of the Normal Stars is named "The Small Star Which is 4 Cubits Behind the King" (MÚL TUR *šá* 4 KÙŠ *ár* LUGAL).

Fig. 4.2 BM 34664 obverse

Following this problematic introductory statement are the first few entries of the Taurus microzodiac scheme. The scheme must have continued in the now lost final column of the tablet. The entries from the scheme are analyzed below in Sect. 4.5.

4.2.2   BM 34664

BM 34664 (Sp. II 147 + Sp. III 551), almost certainly from Babylon, is a fragment from the upper left part of what was originally a two column tablet (Figs. 4.2 and 4.3). Approximately one-third of the height and a little under one-half of the width of the tablet remains, preserving part of the first column on the obverse and the second column on the reverse. The top edge is preserved and only a little is missing on the left edge. A copy by Pinches is published as LBAT 1503 and the tablet has previously been edited and discussed by Rochberg (2004) as her Text C. Pinches' copy interchanges the obverse and reverse and this arrangement was followed by Rochberg. Although the curvature of the tablet is in agreement with Pinches arrangement, it is certain from the content that the arrangement presented here is correct.

Fig. 4.3 BM 34664 reverse

Transliteration

Obv.

1   [*ki-i ina* I]TU.DU₆ UD.15 *ina* UGU MÚL.ME *ár*.ME [*šá* MÚL.ALLA *šamaš* KUR-*ma*]

2   [*ina* UGU M]ÚL *ku-mar šá* MÚL.UD.KA.DUH ŠÚ-*ma* [3 UŠ 20 NINDA *ár*]

3   [MÚL.ME *ár*].ME *šá* MÚL.ALLA *ana ziq-pi* DU *šamaš* K[I.MIN 7-*tú* HA.LA *šá* MÚL.RÍN]

4   [MUL.L]U *šá* MÚL.RÍN BAR KUR *ina* BAR *ina se-rim* U[D.28.KAM MÚL.AŠ.IKU *meš-hu*]

5   [*im-šu*]h ZI 6 UŠ ⌜40⌝ [NINDA *ár* MÚL.ME *ár*.ME *šá* MÚL.ALLA]

6   [*ana zi*]*q-pi* DU-⌜*ma šamaš* KI.MIN⌝ [8-*tú* HA.LA *šá* MÚL.RIN]

7   [MÚL.MÚL] *šá* MÚL.RÍN GU₄ KUR *ina* [GU₄ *ina še-rim* UD.28 KAM x x]

8   [*meš-hu im-šuh*] ZI 10 UŠ *ár* MÚL.ME *ár*.ME [*šá* MÚL.ALLA *ana ziq-pi* DU-*ma*]

9   [*š*]*amaš* KI.MIN 9-*tú* HA.LA *šá* MÚL.R[ÍN MÚL.MAŠ *šá* MÚL.RÍN]

10   [SIG KUR *ina* S]IG *ina še-rim* UD.28.KAM MÚL.SIPA.[ZI.AN.NA *meš-hu im-šuh*]

11   [Z]I 13 UŠ 20 NINDA *ár* <MÚL.ME *ár*.ME *šá*> MÚL.ALLA [*ana ziq-pi*
        DU-*ma šamaš* KI.MIN]
12   [10]-*tú* ḪA.LA *šá* MÚL.RÍN MÚL.ALLA [*šá* MÚL.RÍN ŠU KUR *ina* ŠU
        ina]
13   [*še-rim* U]D.28 MÚL.KAK.SI.SÁ *meš-ḫ*[*u im-šuḫ* ZI 16 UŠ 40 NINDA]
14   [*ár* MÚL.2] MÚL.ME *šá* SAG.DU [MÚL.UR.GU.LA *ana ziq-pi* DU-*ma*
        šamaš KI.MIN]
15   [11-*tú*] ⸢ḪA.LA⸣ *šá* MÚL.RÍN [MÚL.A *šá* MÚL.RÍN IZI KUR *ina* IZI]
16   [*ina še-rim* UD].⸢28⸣ […]

Rev.

1'   […] x x x x […]
2'   [MÚL.EN.TE.N]A.BAR.ḪUM *meš-ḫu im-šuḫ* ZI ⸢6⸣ U[Š 40 NINDA]
3'   ⸢*ár*⸣ MÚL.GAŠAN.TIN *ana ziq-pi* DU-*ma šamaš* KI.MIN 10-*tú* ḪA.LA *šá*
        MÚL A[Š.IKU]
4'   MÚL.PA.BIL.SAG *šá* MÚL.AŠ.IKU GAN KUR *ina* GAN *ina* KIN.SIG U
        [D.30.KAM]
5'   MÚL.LUGAL *meš-ḫu im-šuḫ* ZI 8 UŠ 20 NINDA *ár* MÚL.GA[ŠAN.TIN]
6'   *ana ziq-pi* DU-*ma šamaš* KI.MIN 11-*tú* ḪA.LA *šá* MÚL AŠ.IK[U]
7'   MÚL.MÁŠ *šá* MÚL.AŠ.IKU AB KUR *ina* AB *ina* KIN.SIG UD.30.KAM
        MÚL.Ù[Z?]
8'   *meš-ḫu im-šuḫ* ZI 10 UŠ *ár* MÚL.GAŠAN.TIN *ana ziq-pi* D[U-*ma*]
9'   ⸢*šamaš*⸣ KI.MIN 12-*tú* ḪA.LA *šá* MÚL.AŠ.IKU MÚL.GU *šá* [MÚL.AŠ.
        IKU]
10'   [ZÍ]Z KUR *ina* ZÍZ *ina* KIN.SIG TE(error for: UD).30.KAM MÚL.UGA.
        MUŠEN *me*[*š-ḫu*]
11'   ⸢*im*⸣-*šuḫ* ZI PAP 2/3 DANNA TA MÚL.*dele* EN [10 UŠ]
12'   [*á*]*r* MÚL.GAŠAN.TIN MÚL.AŠ.IKU TA SAG-*šú* EN ⸢TIL⸣-*šú* [KUR-*ḫa*]
13'   ⸢1⸣ UŠ 40 NINDA *ziq-pi i-lak-ma* 2 UŠ 30 NINDA x […]
14'   ⸢x ḪA.LA⸣ *šá* MÚL.AŠ.IKU KUR-*ḫa ina* 12 ḪA.LA.MEŠ 2/3 DAN[NA
        ziq-pi i-lak …]

Translation

Obv.

1   [When in Month VII, day 15, [the Sun rises] at the culmination of The Rear
        Stars [of The Crab, the Sun rises and]
2   sets [at the culmination of] The Shoulder of the Panther. [3 UŠ 20 NINDA
        behind]
3   [The Rear Stars] of the Crab culminate (and) the Sun di[tto (at) the 7th
        portion of Libra]
4   [(which is) Ar]ies of Libra (and corresponds to) Month I rises. In Month I,
        in the morning (which corresponds to) [the 28th day The Field exp]els [a
        flare.]

5   6 UŠ 40 [NINDA behind The Rear Stars of the Crab]
6   culminate and the sun ditto (at) [the 8th portion of Libra]
7   [(which is) Taurus] of Libra (and corresponds to) Month II rises. In [Month
        II, in the morning (which corresponds to) the 28th day …]
8   [exp]els [a flare.] 10 UŠ behind The Rear Stars [of the Crab culminate and]
9   [the S]un ditto (at) the 9th portion of Li[bra (which is) Gemini of Libra]
10   [(and corresponds to) Month III rises. In Month I]II, in the morning (which
        corresponds to) the 28th day The True Shepherd [of Anu expe]ls [a flare.]
11   13 UŠ 20 NINDA behind <The Rear Stars of> the Crab [culminates and
        the Sun ditto]
12   (at) [The 10]th portion of Libra (which is) Cancer [of Libra (and corre-
        sponds to) Month IV rises. In Month IV in]
13   [the morning (which corresponds to)] the 28th day The Arrow [expels a]
        flare. [1 UŠ 40 NINDA]
14   [behind The 2] Stars of the Head of [the Lion culminates and the Sun ditto]
15   [(at) the 11th] portion of Libra [(which is) Leo of Libra (and corresponds
        to) Month V. In Month V,]
16   [in the morning (which corresponds to) the] 28[th day … expels a flare.]

Rev.

1'   […] … […]
2'   [EN.TE.N]A.BAR.ḪUM expels a flare. 6 U[Š 40 NINDA]
3'   behind The Lady of Life culminates and the Sun ditto (at) the 10th portion
        of Pi[sces]
4'   (which is) Sagittarius of Pisces (and corresponds to) Month IX rises. In
        Month IX, in the afternoon (which corresponds to) [the 30th] d[ay]
5'   The King expels a flare. 8 UŠ 20 NINDA behind The Lad[y of Life]
6'   culminates and the Sun ditto (at) the 11th portion of Pisc[es]
7'   (which is) Capricorn of Pisces (and corresponds to) Month X rises. In
        Month X, in the afternoon (which corresponds to) the 30th day The She-
        [Goat]
8'   expels a flare. 10 UŠ behind The Lade of Life culmina[tes and]
9'   the Sun ditto (at) the 12th portion of Pisces (which is) Aquarius of [Pisces]
10'   [(and corresponds to) Month X]I rises. In Month XI in the afternoon (which
        corresponds to) the 30th day The Raven expels a flare.
11'   Total 2/3 bēru. From The Single Star to [10 UŠ]
12'   [behind The Lady of Light, Pisces, from its beginning to its end [rises.]
13'   1 UŠ 40 NINDA culminates and 2 UŠ 30 NINDA … […]
14'   … portion of Pisces rises. In 12 portions 2/3 bēru [culminates.]

Critical Apparatus and Philological Notes

Obv. 5   There appears to be an extra wedge at the bottom right of the ZI sign.
            Pinches copied 5 rather than 6, but collation confirms the reading 6.

Obv. 11   The scribe has omitted the words MÚL.ME *ár*.ME *šá* "The Rear Stars of" in the name of the *Ziqpu* Star.

Obv. 16   Only 24 + x is preserved of the 28.

Rev. 13'   Rochberg read DIRI instead of *ma*?; collation shows that *ma* is more likely.

Commentary

When complete the tablet contained the complete microzodiac rising time scheme for the second half of the zodiac (i.e., from Libra to Pisces), corresponding to the second half of the schematic year. Similar to BM 34713 discussed in the previous section, the tablet begins with a statement of the culminating points at sunrise and sunset on the equinox, this time the autumnal equinox on the 15th day of Month VI. The first section continues by presenting the microzodiac scheme beginning with the 7th portion of Libra through to the 12th and final portion of that zodiacal sign. The reverse preserved the end of the microzodiac scheme for Pisces followed by a short summary of the rising time scheme for that zodiacal sign.

For an analysis of the data contained in the rising time scheme, see Sect. 4.5. The entries for the 7th to the 10th portions of Libra refer to The Rear Stars of the Crab as a *Ziqpu* Star. This star is not listed in the various *Ziqpu*-Star lists, which include only a single entry for The Crab, rather than parts of the constellation. Four stars within the Crab are used as Normal Stars: The Front Star of the Crab to the North (η Cancri), The Front Star of the Crab to the South (θ Cancri), The Rear Star of the Crab to the North (γ Cancri) and The Rear Star of the Crab to the South (δ Cancri). The Rear Stars of the Crab used as *Ziqpu* Stars presumably refers to the two Rear Stars γ and δ Cancri. In BM 36609 + Section 9 the Rear Stars of the Crab are said to be separated from the Front Stars of the Crab by 5 UŠ. Placing the Rear Stars of the Crab 5 UŠ behind The Crab in the *Ziqpu*-Star lists results in agreement between the entries found on this tablet with the reconstruction of the whole scheme presented in Sect. 4.5.

4.2.3   A 3427

A 3427, a tablet from Uruk, formed the starting point for Schaumberger's initial analysis of the microzodiac rising time scheme. Schaumberger (1955) provided a transliteration of the obverse of the tablet and the final line of the colophon and explained the contents. Rochberg (2004) subsequently transliterated and translated the obverse of the tablet and discussed its content in detail.

A 3427 is a substantial fragment preserving about half the width and probably a little over half the height of the original tablet (Figs. 4.4, 4.5 and 4.6). The obverse is well preserved and clearly written, but the reverse has been largely obliterated with only a few legible signs remaining. The tablet is divided into sections by horizontal rulings. Each section contains the microzodiac rising time scheme for one sign of the zodiac. When complete, the tablet covered the signs Scorpio,

Fig. 4.4 A 3427 *upper* edge (courtesy of the Oriental Institute of the University of Chicago)

Fig. 4.5 A 3427 obverse
(courtesy of the Oriental
Institute of the University of
Chicago)

Sagittarius, Capricorn and Aquarius. My transliteration of the obverse is largely
based upon Rochberg's edition, although I have checked readings against a photograph where necessary.

Transliteration

Upper edge

1 *ina a-mat* ᵈ60 *u* ᵈ*An-tum liš-lim*

Fig. 4.6 A 3427 reverse
(courtesy of the Oriental
Institute of the University of
Chicago)

Obv.

1   [T]A 5 UŠ *ár* 2 MÚL.ME *šá* SAG MÚL.A EN 5 UŠ *ár* MÚL.DELE *šá*
     KUN-*šú* MÚL.[GÍR.TAB TA SAG-*šú* EN TIL-*šú* KUR]

2   8 UŠ 20 NINDA *ár* 2 MÚL.ME *šá* SAG ʌ KI.MIN ḪA.LA *reš-tú šá*
     MÚL.GÍR.TAB MÚL.GÍR.TAB *š*[*á* MÚL.GÍR.TAB ITU.APIN KUR *ina*
     ITU.APIN *ina še-rim* UD.28.KAM]

3   MÚL.UR.IDIM *meš-ḫa im-šuḫ* ZI 1 UŠ 40 NINDA *ár* 4 *šá* GABA-*šú* KI.
     MIN 2-*tú* Ḫ[A.LA *šá* MÚL.GÍR.TAB MÚL.PA *šá* MÚL.GÍR.TAB ITU.
     GAN KUR]

4   *ina* ITU.GAN *ina še-rim* UD.28 ᵈ*ṣal-bat-a-nu* KI.MIN 5 UŠ *ár* 4 *šá*
     GABA-*šú* [KI.MIN 3-*tú* ḪA.LA *šá* MÚL.GÍR.TAB MÚL.MÁŠ *šá* MÚL.
     GÍR.TAB]

5   ITU.AB KUR *ina* ITU.AB *ina še-rim* UD.28 MÚL.ALLA KI.MIN 8 UŠ
     20 NINDA *ár* 4 *šá* ꞌGABA-*šúꞌ* [KI.MIN 4-*tú* ḪA.LA *šá* MÚL.GÍR.TAB
     MÚL.GU *šá* MÚL.GÍR.TAB]

6   ITU.ZÍZ KUR *ina* ITU.ZÍZ *ina še-rim* UD.28 MÚL.*nu-muš-da* KI.MIN 11
     UŠ 40 NINDA *ár* 4 [*šá* GABA-*šú* KI.MIN 5-*tú* ḪA.LA *šá* MÚL.GÍR.TAB
     MÚL.AŠ.IKU *šá* MÚL.GÍR.TAB]

7   ITU.ŠE KUR *ina* ITU.ŠE *ina še-rim* UD.28 MÚL.KU$_6$ KI.MIN ½
     DANNA *ár* 4 *šá* GABA-š[*ú* KI.MIN 6-*tú* ḪA.LA *šá* MÚL.GÍR.TAB
     MÚL.LU *šá* MÚL.GÍR.TAB]

8   ITU.BAR KUR *ina* ITU.BAR *ina še-rim* UD.28 <MÚL> .AŠ.IKU KI.
     MIN 18 UŠ 20 NINDA *ár* 4 *šá* GA[BA-*šú* KI.MIN 7-*tú* ḪA.LA *šá* MÚL.
     GÍR.TAB MÚL.MÚL *šá* MÚL.GÍR.TAB]

9   ITU.GU$_4$ KUR *ina* ITU.GU$_4$ *ina še-rim* UD.28 MÚL.MÚL KI.MIN 1 UŠ
     ⌜40⌝ NINDA *ár* 2 *šá* GIŠ.K[UN-*šú* KI.MIN 8-*tú* ḪA.LA *šá* MÚL.GÍR.
     TAB MÚL.MAŠ.MAŠ *šá* MÚL.GÍR.TAB]

10   ITU.SIG KUR *ina* ITU.SIG *ina še-rim* UD.28 SIPA KI.MIN 5 UŠ *ár* 2 *šá*
     GIŠ.KUN-[*šú* KI.MIN 9-*tú* ḪA.LA *šá* MÚL.GÍR.TAB MÚL.ALLA *šá*
     MÚL.GÍR.TAB]

11   ⌜ITU.ŠU⌝ KUR *ina* ITU.ŠU *ina še-rim* UD.28 MÚL.KAK.SI.SÁ KI.MIN 8
     UŠ 20 NINDA *ár* 2 *šá* GIŠ.[KUN-*šú* KI.MIN 10-*tú* ḪA.LA *šá* MÚL.GÍR.
     TAB MÚL.A *šá* MÚL.GÍR.TAB]

12   ⌜ITU.IZI⌝ KUR *ina* ITU.IZI *ina še-rim* UD.28 MÚL.BAN KI.MIN 1 UŠ 40
     NINDA *ár* MUL.DELE *šá* K[UN-*šú* KI.MIN 11-*tú* ḪA.LA *šá* MÚL.GÍR.
     TAB MÚL.ABSIN *šá* MÚL.GÍR.TAB]

13   ITU.KIN KUR *ina* ITU.KIN *ina še-rim* UD.28 MÚL.BIR KI.MIN 5 UŠ *ár*
     MÚL.DELE *šá* KUN-[*šú* KI.MIN 12-*tú* ḪA.LA *šá* MÚL.GÍR.TAB MÚL.
     RÍN *šá* MÚL.GÍR.TAB]

14   ITU.DU$_6$ KUR *ina* ITU.DU$_6$ *ina še-rim* UD.28 MÚL.NIN.MAḪ KI.MIN
     PAP 1 DANNA 10 UŠ [TA 5 UŠ *ár* 2 MÚL.ME *šá* SAG MÚL.A EN 5
     UŠ]

15   *ár* MÚL.DELE *šá* KUN-*šú* MÚL.GÍR.TAB TA SAG-*šú* EN TIL-*šú* KUR-
     ḫa 1-*et* ḪA.LA [3 UŠ 20 NINDA *ziq-pi i-lak-ma*]

16   2 UŠ 30 NINDA 1-*et* ḪA.LA *šá* MÚL.GÍR.TAB KUR-*ḫa ina* 12 ḪA.LA 1
     DANNA 10 U[Š *ziq-pi i-lak-ma* MÚL.GÍR.TAB]

17   TA SAG-*šú* EN TIL-*šú* KUR-*ḫa* PAP 2 DANNA *ina ziq-pi i-lak-ma* MÚL.
     [... TA ...]

18   EN TIL-*šú* 1 ½ DANNA KUR NIM.MA SAR
     -
19   TA 5 UŠ *ár* MÚL.DELE *šá* KUN-*šú* EN MÚL.*na-at-tíl-lum* MÚL.PA TA
     SAG-*šú* [EN TIL-*šú* KUR 8 UŠ 20 NINDA *ár* MÚL.DELE *šá* KUN-*šú*]

20   *ana ziq-pi* DU-*ma* ᵈUTU KI.MIN ḪA.LA *reš-tú šá* SAG MÚL.PA *šá*
     MÚL.PA IT[U.GAN KUR *ina* ITU.GAN *ina* AN.NE UD.29]

21   [M]ÚL.GÍR.TAB *meš-ḫi im-šuḫ* ZI 8 UŠ (error for: 1 UŠ 20 NINDA) *ár*
     MÚL.*e₄-ru₆* KI.MIN 2-*tú* ḪA.L[A *šá* MÚL.PA MÚL.MÁŠ *šá* MÚL.PA
     ITU.AB KUR]

22   [*ina*] ITU.AB *ina* AN.NE UD.29 MÚL.UD.KA.DUḪ.A KI.MIN 8 (error
     for: 5) UŠ *ár* MÚL.*e₄-ru₆* [KI.MIN 3-*tú šá* ḪA.LA MÚL.PA MÚL.GU *šá*
     MÚL.PA ITU.ZÍZ KUR]

23   [*ina*] ITU.ZÍZ *ina* AN.NE UD.29 MÚL.ALLA KI.MIN 8 UŠ 20 NINDA
     ár MÚL.*e₄*-⌜*ru₆*⌝ [KI.MIN 4-*tú šá* ḪA.LA MÚL.PA MÚL.AŠ.IKU *šá*
     MÚL.PA ITU.ŠE KUR]

24   [*ina*] ITU.ŠE *ina* AN.NE UD 29 MÚL.NIN.MAḪ KI.MIN 11 UŠ 40
       NINDA *ár* MÚL.*e₄-ru₆* K[I.MIN 5-*tú šá* ḪA.LA MÚL.PA MÚL.LU *šá*
       MÚL.PA ITU.BAR KUR]

25   [*ina*] ⌜ITU.BAR⌝ *ina* AN.NE UD.29 MÚL⌝.KA₅.A KI.MIN ⌜½ DANNA *ár*
       MÚL.*e₄-ru₆* KI⌝.[MIN 6-*tú šá* ḪA.LA MÚL.PA MÚL.MÚL *šá* MÚL.PA
       ITU.GU₄ KUR]

Rev.

1'   […] x […]
2'   [… KI].⌜MIN⌝ 5 ⌜ḪA⌝.LA […]
3'   […] ⌜KI.MIN⌝ 6 ḪA.⌜LA⌝ […]
4'   […] ⌜KI.MIN⌝ 7 ḪA.LA ⌜*šá*⌝ […]
5'   [… MÚL.*tak.šá*]-⌜*a-tú* KI.MIN 8 ḪA⌝.LA […]
6'   [… MÚL.*tak.šá*]-⌜*a-tú* KI⌝.MIN ⌜9⌝ ḪA.⌜LA⌝ […]
7'   [… MÚL.*tak.šá-a*]-⌜*tú* KI⌝.MIN 10 ⌜ḪA.LA⌝ […]
8'   [… MÚL.*tak.šá-a*]-⌜*tú* KI.MIN 11 ḪA⌝.[LA …]
9'   [… KI.MIN] ⌜12⌝ [ḪA.LA …]
10'   […] ⌜x x⌝ […]
11'   […] ⌜x x⌝ […]
12'   […] EN AB? ⌜x x x⌝ […]

- -

13'   […] ⌜x x x⌝ […]
14'   […]⌜x x x x x x x x x x x⌝ […]
15'   ⌜x x x x x x x x⌝ ᵈ1 *u* ⌜x x x⌝ […]
16'   ⌜TIR⌝.AN.NAᵏⁱ-*ú pa-liḫ* ᵈ1 ᵈEN.LÍL *u* ᵈ*é-a* NU TÙM-*šú ina* […]

Translation

Upper edge

1   At the command of Anu and Antu may it go well.

Obv.

1   [Fr]om 5 UŠ behind The 2 Stars of the Head of the Lion to 5 UŠ behind
      The Single Star of its Tail, S[corpio from its beginning to its end rises].

2   8 UŠ 20 NINDA behind The 2 Stars of the Head of the Lion ditto (at) the
      first portion of Scorpio (which is) Scorpio o[f Scorpio (and corresponds to)
      Month VIII rises. In Month VIII, in the morning (which corresponds to) the
      28th day,]

3   The Wolf expels a flare. 1 UŠ 40 NINDA behind The 4 Stars of its Breast
      ditto (at) the 2nd por[tion of Scorpio (which is) Sagittarius of Scorpio (and
      corresponds to) Month IX rises.]

4   In Month IX, in the morning (which corresponds to) day 28, Mars ditto.
      5 UŠ behind The 4 Stars of its Breast [ditto (at) the 3rd portion of Scorpio
      (which is) Capricorn of Scorpio]

5   (and corresponds to) Month X rises. In Month X, in the morning (which
         corresponds to) day 28, The Crab ditto. 8 UŠ 20 NINDA behind The 4
         Stars of its Breast [ditto (at) the 4th portion of Scorpio (which is) Aquarius
         of Scorpio]

6   (and corresponds to) Month XI rises. In Month XI, in the morning (which
         corresponds to) day 28, Numušda ditto. 11 UŠ 40 NINDA behind The 4
         [Stars of its Breast ditto (at) the 5th portion of Scorpio (which is) Pisces of
         Scorpio]

7   (and corresponds to) Month XII rises. In Month XII, in the morning (which
         corresponds to day 28, The Fish ditto. ½ bēru behind The 4 Stars of it[s]
         Breast [ditto (at) the 6th portion of Scorpio (which is) Aries of Scorpio]

8   (and corresponds to) Month I rises. In Month I, in the morning, day 28, The
         Field ditto. 18 UŠ 20 NINDA behind The 4 Stars of [its] Bre[ast ditto
         (at) the 7th portion of Scorpio (which is) Taurus of Scorpio]

9   (and corresponds to) Month II rises. In Month II, in the morning (which
         corresponds to) day 28, The Stars ditto. 1 UŠ 40 NINDA behind The 2
         Stars of [its] Ru[mp ditto (at) the 8th portion of Scorpio (which is) Gemini
         of Scorpio]

10   (and which corresponds to) Month III rises. In Month III, in the morning
         (which corresponds to) day 28, The True Shepherd ditto. 5 UŠ behind The
         2 Stars of [its] Rump [ditto (at) the 9th portion of Scorpio (which is) Cancer
         of Scorpio]

11   (and corresponds to) Month IV rises. In Month IV, in the morning (which
         corresponds to) day 28, The Arrow ditto. 8 UŠ 20 NINDA behind The 2
         Stars of [its] Ru[mp ditto (at) the 10th portion of Scorpio (which is) Leo of
         Scorpio]

12   (and corresponds to) Month V rises. In Month V, in the morning (which
         corresponds to) day 28, The Bow ditto. 1 UŠ 40 NINDA behind The Single
         Star of its Ta[il ditto (at) the 11th portion of Scorpio (which is) Virgo of
         Scorpio]

13   (and corresponds to) Month VI rises. In Month VI, in the morning (which
         corresponds to) day 28, The Kidney ditto. 5 UŠ behind The Single Star of
         [its] Tail [ditto (at) the 12th portion of Scorpio (which is) Libra of Scorpio]

14   (and corresponds to) Month VII rises. In Month VII, in the morning (which
         corresponds to) day 28, Ninmaḫ ditto. Total 1 bēru 10 UŠ [from 5 UŠ
         behind The 2 Stars of the Head of the Lion to 5 UŠ]

15   behind The Single Star of its Tail, Scorpio, from its beginning to its end,
         rises. 1 portion [3 UŠ 20 NINDA culminates and]

16   2 UŠ 30 NINDA (which is) 1 portion of Scorpio rises. In 12 portions, 1
         bēru 10 U[Š culminates and Scorpio]

17   from its beginning to its end rises. Total 2 bēru culminates and [… from …]

18   to its end 1 ½ bēru rises in the east and becomes visible.
- -
19   From 5 UŠ behind The Single Star if its Tail to The Harness (error for: The
         Yoke?) Sagittarius, from its beginning [to its its end rises. 8 UŠ 20 NINDA
         behind The Single Star of its Tail]

20   culminate and the Sun ditto (at) the first portion of the beginning (which is)
        Sagittarius of Sagittarius (and corresponds to) Mon[th X rises. In Month X,
        in the noontime (which corresponds to) day 29]
21   Scorpio expels a flare. 8 UŠ (error for: 1 UŠ 20 NINDA) behind Eru ditto
        (at) the 2nd porti[on of Sagittarius (which is) Capricorn of Sagittarius (and
        corresponds to) Month X rises.]
22   [In] Month X, in the noontime (which corresponds to) day 29, The Panther
        ditto. 8 (error for: 5) UŠ behind Eru [ditto (at) the 3rd portion of Sagittarius
        (which is) Aquarius of Sagittarius (and corresponds to) Month XI rises.]
23   [In] Month XI, in the noontime (which corresponds to) day 29, The Crab
        ditto. 8 UŠ 20 NINDA behind Eru [ditto (at) the 4th portion of Sagittarius
        (which is) Pisces of Sagittarius (and corresponds to) Month XII rises.]
24   [In] Month XII, in the noontime (which corresponds to) day 29, Ninmaḫ
        ditto. 11 UŠ 40 NINDA behind Eru di[tto (at) the 5th portion of Sagittarius
        (which is) Aries of Sagittarius (and corresponds to) Month I rises.]
25   [In] Month I, in the noontime (which corresponds to) day 29, The Fox ditto.
        ½ bēru behind Eru dit[to (at) the 6th portion of Sagittarius (which is)
        Taurus of Sagittarius (and corresponds to) Month II rises.]

Rev.

1'   [...] x [...]
2'   [... di]tto (at) the 5th portion [...]
3'   [...] ditto (at) the 6th portion [...]
4'   [...] ditto (at) the 7th portion of [...]
5'   [... The Star from the Trip]lets ditto (at) the 8th portion [...]
6'   [... The Star from the Trip]lets ditto (at) the 9th portion [...]
7'   [... The Star from the Triple]ts ditto (at) the 10th portion [...]
8'   [... The Star from the Triple]ts ditto (at) the 11th port[ion ...]
9'   [... ditto] (at) the 12th [portion ...]
10'   [...] ... [...]
11'   [...] ... [...]
12'   [...] to ... [...]
- -
13'   [...] ... [...]
14'   [...] ... [...]
15'   ... [...]
16'   Tiranean. Whoever reveres An, Antu and Ea shall not take it away. [...]

Critical Apparatus and Philological Notes

Obv. 2   Here and throughout the text the scribe uses the ditto mark KI.MIN
                in place of a reference to culmination. This suggests that the text was
                part of a series and that this passage was written out in full only on
                the first tablet of the series.
Obv. 4   Here and throughout the remainder of the text the scribe uses the ditto
                mark KI.MIN to replace the phrase *meš-ḫu im-šuḫ* ZI, which is given

	in Obv. 3. Inclusion of the ZI within the ditto provides the strongest evidence that this sign is to be taken as part of the statement *meš-ḫu im-šuḫ* ZI and not as the beginning of the next statement.
Obv. 9–13	As noted already by Brown (2005: 413), Rochberg wrongly "corrects" the culminating position in her translation to <2>1;40, <2>5, etc.
Obv. 19	The scribe appears to have mistakenly given the name of the *Ziqpu* Star, The Harness (MÚL.*na-at-tíl-lum*) instead of the next *Ziqpu* Star, The Yoke (MÚL.*ni-i-ri* or MÚL.ŠUDUN).
Obv. 21	8 UŠ is a clear scribal error for 1 UŠ 40 NINDA.
Obv. 22	Again, the scribe mistakenly writes 8 UŠ, this time instead of 5 UŠ. This mistake is more understandable as the signs 8 and 5 are similar.
Rev. 5'	Only the final signs *-a-tú* are preserved of the *Ziqpu* Star name. This identifies the star as either MÚL *šá maš-a-tú* or MÚL *šá tak-šá-a-tú*. Both stars appear only in the microzodiac rising time scheme for the zodiacal sign Aquarius, thus identifying this section. The latter may be restored here on the basis of the reconstructed scheme discussed in Sect. 4.5.
Rev. 13'–15'	The first three lines of the colophon are very badly damaged and will require careful collation of the tablet in order to be read.

Commentary

The obverse contains a copy of the microzodiac rising time scheme for Scorpio and the beginning of Sagittarius. The text is somewhat more abbreviated, making more use of the ditto marker KI.MIN than BM 34713 and BM 34664, but essentially follows the same pattern in its entries.

Like the other tablets, the sections begin and end with a summary of the scheme for that sign of the zodiac. The beginning summary is very short, simply giving the range of culminating points between the rising of the beginning and end of the zodiacal sign. The summary at the end of the section, however, follows BM 34713 and BM 34664 is stating that the range in culminating points is 3 UŠ 20 NINDA for 1 microzodiac portion of 2 UŠ 30 NINDA, and then that 1 *bēru* 10 UŠ culminates for the 30 UŠ of the zodiacal sign of Scorpio. However, the text continues with the statement that 2 *bēru* culmination corresponds to a range that ends with the end of Scorpio and which should, according to the mathematics of the scheme, be 1½ *bēru* or 1½ signs of the zodiac. Unfortunately, the tablet is broken at the crucial point and so it is not certain where the range begins. Two possibilities exist: first, the range could begin in the middle of the previous sign, which is Libra, and extend to the end of Scorpio. The middle of Libra is the midpoint of the microzodiac scheme which, from the long introduction in BM 34713, seems to begin at the midpoint of Aries. Alternatively, the 2 *bēru* could be a simple scribal error.[6]

---

[6]Schaumberger (1955: 241) considered the possibility that the text refers to zodiacal sign of Scorpio which includes the rear balance of Libra as part of the constellation. There is no support for this suggestion in Babylonian sources.

The second section presents the scheme for Sagittarius. As noted by Schaumberger (1955: 241) and Rochberg (2004), the text contains a major scribal error in the entries for the 2nd and 3rd portions: in both cases the scribe has written that it is the point 8 UŠ behind Eru which culminates when he should have written 1 UŠ 40 NINDA and 5 UŠ respectively. The summary statement at the beginning of the section almost certainly also contains a scribal error. The text as written states that the rising of Sagittarius from its beginning to its end corresponds to the culmination of the range from 5 UŠ behind The Single Star if its Tail to The Harness. However, the Harness is only 30 UŠ behind the Single Star of its Tail. It seems most likely that the scribe has written the wrong star name here, giving The Harness instead of the following Ziqpu Star, The Yoke. An alternative possibility is that he omitted the words "10 UŠ behind" before naming The Harness.

The reverse of the tablet is very badly damaged, and only a few signs can be read. Nevertheless, sufficient traces remain to identify that the preserved section concerns Aquarius. The text ends with a badly preserved colophon.

4.3 Texts Presenting the Simplified Microzodic Rising Time Scheme

Three tablets contain texts which present a simplified version of the microzodiac scheme in which only the culminating point corresponding to the microzodiac sign is given, with no statements about the corresponding month, day number or the expelling of a flare by a star.

4.3.1 BM 35456

BM 35456 (Sp. II, 1045) is a small fragment from the bottom right corner of a tablet (Figs. 4.7, 4.8, 4.9 and 4.10). A copy of the tablet by Pinches is published as LBAT 1505. Obverse and reverse should be interchanged in Pinches' copy.

The preserved part of the tablet concerns the zodiacal signs of Virgo (Obv. and lower edge) and Libra (reverse). Assuming that the tablet was divided into two columns on each of the obverse and reverse, it is quite possible that the whole tablet covered all twelve signs of the zodiac, six on each of the obverse and reverse beginning with Aries, giving the complete microzodiac scheme.

Transliteration

Obv.

1′   [... 8 U]Š 20 NINDA *ár rit-tu₄* GÀM
2′   [... 11 U]Š 40 NINDA *ár rit-tu₄* GÀM

Fig. 4.7 BM 35456 obverse

Fig. 4.8 BM 35456 *lower* edge

Fig. 4.9 BM 35456 reverse

Fig. 4.10 BM 35456 *right* edge

3' [... 1/2] DANNA *ár rit-tu₄* GÀM
4' [...] ⌜1⌝8 UŠ 20 NINDA *ár rit-tu₄* GÀM
5' [...] ⌜2⌝1 UŠ 40 NINDA *ár* ⌜*rit-tu₄* GÀM⌝

Lower Edge

1 [...] MÚL.MAŠ.MAŠ IGI-⌜*i ana ziq-pi*⌝
2 [...] ⌜3⌝ UŠ 20 NINDA *ár* MAŠ.MAŠ IGI-⌜*i*⌝
3 [...] ⌜1⌝ UŠ 40 NINDA *ár* MAŠ.MAŠ *ár-ki-i*
4 [...] ⌜5⌝ UŠ *ár* MAŠ.MAŠ *ár-ki-*⌜*i*⌝

Rev.

1 [T]A 5 UŠ *ár* MAŠ.MAŠ *ár-ki-i*
2 [E]N 5 UŠ *ár* 2 MÚL.MEŠ *šá* SAG MÚL.A
3 RÍN TA SAG-*šú* EN TIL-*šú* 30 UŠ KUR? u 40 ⌜x⌝-*ú*
4 [... 8 U]Š 20 NINDA *ár* MÚL.MAŠ.MAŠ *ár-ki-i ana ziq-pi* DU-*ma*
5 [... 11 UŠ] ⌜40⌝ NINDA *ár* MÚL.MAŠ.MAŠ *ár*⌝-*ki-*⌜*i*⌝ [...]

Translation

Obv.

1' [... 8 U]Š 20 NINDA behind The Handle of the Crook.
2' [... 11 U]Š 40 NINDA behind The Handle of the Crook.
3' [... 1/2] *bēru* behind The Handle of the Crook.
4' [...] 18 UŠ 20 NINDA behind The Handle of the Crook.
5' [...] 21 UŠ 40 NINDA behind The Handle of the Crook.

Lower Edge

1	[...] The Front Twin culminates.
2	[...] ⌜3⌝ UŠ 20 NINDA behind The Front Twin.
3	[...] ⌜1⌝ UŠ 40 NINDA behind The Rear Twin.
4	[...] ⌜5⌝ UŠ behind The Rear Twin.

- -

Rev.

1	[Fr]om 5 UŠ behind The Rear Twin
2	[t]o 5 UŠ behind The Two Stars of the Head of the Lion.
3	Libra, from its beginning to its end 30 UŠ rises and 40 [...]
4	[... 8 U]Š 20 NINDA behind The Rear Twin culminates.
5	[... 11 UŠ] ⌜40 NINDA behind The Rear Twin [...].

Critical Apparatus and Philological Notes

Rev. 1   Pinches' copy has ZA instead of 5, but 5 is clear from collation.
Rev. 3   The signs copied by Pinches as GABA MEŠ 30 are to be read as UŠ KUR$^?$
            u 40. I cannot read the following damaged sign. The *ú* sign is written
            below this damaged final sign when the scribe ran out of room on the line.

Commentary

The text is divided into sections each of which concerns one sign of the zodiac. Parts of the sections for Virgo and Libra are preserved; when complete the tablet probably contained sections for all twelve signs of the zodiac.

The sections begin with a statement of the range of culminating points which correspond to the rising of the complete sign of the zodiac. This statement is followed by twelve one-line statements giving the culminating points for each of the twelve microzodiac signs within the sign of the zodiac. The identification of the individual microzodiac signs is lost in the missing text at the beginning of each line, but it is clear that Obv. 1′ to lower edge 4 are microzodiac signs number 4–12 of Virgo and Rev. 4–5 are the first two microzodiac signs of Libra.

This text attests to a difference between the standard *Ziqpu* Star list and the *Ziqpu* Stars as used in the microzodiac rising time scheme. In both the Virgo and Libra sections of this tablet the position of the culminating point increases by 3 UŠ 20 NINDA at successive microzodiac signs. The entries in lines Obv. 5′ and Lower Edge 1 give the culminating position as 21 UŠ 40 NINDA behind The Handle of the Crook and at The Front Twin Star, which implies that The Front Twin Star is 25 UŠ behind The Handle of the Crook. Lower Edge 2 and 3 imply that The Rear Twin Star is 5 UŠ behind The Front Twin Star. The standard 25-star *Ziqpu* Star list gives only one position for The Twins, placing them 30 UŠ behind The Handle of the Crook. The 26-star *Ziqpu* Star list and BM 36609 + Sect. 9 both indicate that The Rear Twin Star is 5 UŠ behind The Front Twin Star. However, in the 26-star *Ziqpu* Star list, the distance between The Handle of the Crook and The Twins is also given as 30 UŠ. In both the standard 25- star and the 26-star list The Crab is placed 20 UŠ behind the Twins. I have previously assumed that the distance to The Crab should therefore be interpreted 20 UŠ behind The Front Twin, and therefore 15 UŠ behind The Rear Twin. The current text implies, however, that The Twins in the standard 25-star *Ziqpu* Star list should be taken as the position of The Rear Twin, and that The Crab is 20 UŠ behind The Rear Twin, and also that The Front Twin is 25 UŠ rather than 30 UŠ behind The Handle of the Crook whereas The Rear Twin is 30 UŠ behind The Handle of the Crook.

A full discussion of the microzodiac data preserved in this tablet may be found in Sect. 4.5.

4.3.2   BM 46167

BM 46167 (81–7–6, 628) is a small flake from Babylon containing a few lines from a microzodiac rising time scheme (Fig. 4.11). Although it is hard to be certain from such a small fragment, the absence of any references to months, days, or stars expelling a flare suggests that it is from the simplified microzodiac scheme.

Transliteration

1′   [...] *ár* ⸢MÚL *šá tak*?⸣-[*šá-a-tum* ...]
2′   [... *-t*]*i* ḪA.LA *šá* [...]
3′   [... 1],⸢40⸣ NINDA *ár* MÚL *e-*[*du* ...]
4′   [... *-t*]*i* ḪA.LA *šá* M[ÚL ...]
5′   [...] MÚL *e-du* [...]
6′   [...] ⸢ḪA.LA?⸣ [...]

Translation

1′   [...] behind The Star from the Tri[plets ...]
2′   [...] portion of [...]
3′   [... 1],⸢40⸣ NINDA behind the Sin[gle Star ...]
4′   [...] portion of [...]
5′   [...] The Single Star [...]
6′   [...] portion? [...]

Critical Apparatus and Philological Notes

6′   The reading ḪA.LA is consistent with the traces which remain, but is not certain.

Commentary

This small fragment preserves part of the microzodiac scheme for Aquarius and Pisces. A discussion of the data preserved in this tablet may be found in Sect. 4.5.

Fig. 4.11 BM 46167

4.3.3   *BM 77242*

BM 77242 (93–6–30, 22) is a small flake almost certainly from Babylon
(Fig. 4.12). It was identified as relating to the microzodiac scheme and published by
Horowitz (1994). Collation of the tablet and comparison with other microzodiac
texts allows a slight improvement on Horowitz's edition and also for the position of
the tablet within the microzodiac scheme to be established.

Transliteration

1′   [...] *ár* ⌜4⌝ [MÚL *šá* SI MÚL.LU.LIM ...]
2′   [...] ḪA.LA *šá* [MÚL.AL.LUL ...]
3′   [...] 40 GAR *ár* 4 ⌜MÚL⌝ [*šá* SI MÚL.LU.LIM ...]
4′   [... *-i-t*]*i* ḪA.LA *šá* MÚL.[AL.LUL ...]
5′   [... U]Š *ár* 4 MÚL *šá* SI MÚ[L.LU.LIM ...]
6′   [...]-*i-ti* ḪA.LA *šá* MÚL.[AL.LUL ...]
7′   [... 1]⌜3⌝ UŠ 20 GAR *ár* 4 M[ÚL ...]
8′   [...]-⌜*i-ti*⌝ ḪA.LA [*šá* MÚL.AL.LUL ...]
9′   [...] ⌜ḪA.LA⌝ [*šá* MÚL.AL.LUL ...]

Fig. 4.12 BM 77242

Translation

1'   [...] behind The 4 [Stars of the Horn of the Stag ...]
2'   [...] portion [of Cancer ...]
3'   [...6 UŠ] 40 NINDA behind The 4 Stars [of the Horn of the Stag ...]
4'   [...] portion of [Cancer ...]
5'   [... 10 U]Š behind The 4 Stars of the Horn of the [Stag ...]
6'   [...] portion of [Cancer ...]
7'   [... 1]3 UŠ 20 NINDA behind The 4 Sta[rs of the Horn of the Stag ...]
8'   [...] portion [of Cancer ...]
9'   [...] portion [of Cancer ...]

Critical Apparatus and Philological Notes

1'   Horowitz read *šá* instead of 4. The sign is damaged but collation suggests that 4 is the more likely reading.

2'–3' Horowitz marked a ruling between lines 2' and 3' but I cannot see one and none is expected by the context.

3'   Horowitz read *šá* instead of 4, but 4 is clear.

6'   Horowitz restored MÚL.[UR.GU.LA] at the end of this line, but the scheme makes it certain that we are dealing with Cancer, not Leo.

7'   Two vertical wedges are visible at the beginning of the line indicating that the number must end in either 2 or 3.

Commentary
This small fragment preserved entries for the culminating points corresponding to
the microzodiac signs within Cancer. A discussion of the data preserved in this
tablet may be found in Sect. 4.5.

4.3.4   U 196

The tablet U 196 was excavated from the site of the Bīt Reš sanctuary in Uruk.
Schaumberger transliterated part of one side of the tablet from a photograph and
commented upon it, claiming that the tablet attested to a rising arc of 27° for
Capricorn, as is found in System B. However, closer examination of the contents of
the tablet show that it follows the usual microzodiac scheme and that the section
Schaumberger transliterated refers to Sagittarius. The tablet will be edited by
Christine Proust and I in our forthcoming edition of the astronomical and related
cuneiform tablet from Uruk in the Istanbul Museum of Archaeology (Steele and
Proust forthcoming). I therefore provide only a brief overview of the tablet here.

The surface of the tablet is very badly damaged but some text on each side can
be read. The side read by Schaumberger is probably the reverse. It preserves a
single section presenting the microzodiac scheme for Sagittarius with a summary of
the scheme at the end. Unfortunately, the scribe clearly made some mistakes in
writing the text. For example, the scribe seems to have skipped on to the *Ziqpu* Star
The Harness a couple of entries too early before correcting himself. The other side
of the tablet, which is probably the obverse, contains the end of a section presumably dealing with Taurus followed by a section concerning Gemini.

Because so few entries from the microzodiac scheme can be read with confidence, this tablet it is of little use in attempting to analyse the scheme. It is worth
noting, at least, that the entries in the Gemini section, which is better preserved than
the Sagittarius section, agree with those in the scheme as reconstructed in Sect. 4.5.

4.4   A Text Containing Statements of the Rising Times
for Complete Signs of the Zodiac

The microzodiac texts usually include a statement at the beginning and the end of a
section setting out culminating points for the rising of the beginning and end of that
culminating. The tablet BM 36609+ contains a collection of these statements
originally covering the complete twelve signs of the zodiac.

4.4.1 BM 36609+ Obv. III 14–29

BM 36609+ (80–6–17, 339+) is a large fragment from Babylon containing a compendium of texts concerning *Ziqpu* Stars and Normal Stars. The tablet has been edited and studied by Roughton et al. (2004). Section 7 (Obv. III 14–29) contains a list of statements of the rising times for the signs of the zodiac. Other sections of this text also concern the rising time scheme; this material will be discussed in Chap. 5. Unfortunately, entries for only five zodiacal signs (Aries to Leo) are preserved.

Transliteration

Obv. III

14   TA 10 U[Š *ár* M]ÚL.ÙZ
15   EN SA₄ [*šá* GA]BA-*šú*
16   MÚL.ḪUN.GÁ TA ⌜SAG-*šú*⌝ [EN TIL-*š*]*ú* KUR
17   TA SA₄ *šá* GABA-*šú*
18   EN 5 UŠ *ina* IGI *kin-ṣa*
19   MÚL.MÚL TA ⌜SAG⌝-*šú* EN TIL-*šú* KUR
20   TA 5 UŠ ⌜*kin*⌝-*ṣa* NU KUR
21   EN *a-si-du*
22   MAŠ.MAŠ TA SAG-*šú* EN TIL-*šú* KUR
23   TA *a-si-du*
24   EN 5 UŠ *ár um-mu-lut*
25   AL.[LA] ⌜TA SAG⌝-*šú* EN TIL-*šú*
26   ⌜TA⌝ 5 [U]Š *ár um-mu-lut*
27   [E]N 5 UŠ *ár* GÀM
28   [A] ⌜TA SAG⌝-*šú* EN TIL-*šú*
29   [TA 5 UŠ *ár* GÀM]

Translation

14   From 10 U[Š behind] The She-Goat
15   to The Bright Star [of] his [Bre]ast,
16   Aries, from its beginning to its end, rises.
17   From The Bright Star of his Breast
18   to 5 UŠ in front of The Knee,
19   Taurus, from its beginning to its end, rises.
20   From 5 UŠ not reaching The Knee
21   to The Heel,
22   Gemini, from its beginning to its end, rises.
23   From The Heal
24   to 5 UŠ behind The Dusky Stars,
25   Can[cer], from its beginning to its end, rises.

26   From 5 [U]Š behind The Dusky Stars
27   [t]o 5 UŠ behind The Crook,
28   [Leo,] from its beginning to its end, rises.
29   [From 5 UŠ behind] The Crook

Critical Apparatus and Philological Notes

14   The star MÚL.ÙZ "The She-Goat" is almost certainly the same star as
        MÚL.GAŠAN.TIN "The Lady of Life". See the discussion in Roughton
        et al. (2004: 549).
19   The phrase 5 UŠ ⌜kin⌝-ṣa NU KUR "5 UŠ not reaching the Knee" is clearly
        referring to the same position as 5 UŠ *ina* IGI *kin-ṣa* "5 UŠ in front of The
        Knee" found in the previous line. The terminology NU KUR to indicate a
        position in front of a star is also attested in the so-called Atypical Text E
        (BM 41004); see Brack-Bernsen and Hunger (2005–2006). BM 34713
        Rev. I 31 gives the end of the rising arc as 4 UŠ *ina* IGI MÚL.*kin-si*. Both
        entries are confusing as assuming the scheme follows the *Ziqpu* Star lists
        the end of the rising time for Taurus should be exactly at The Knee. See
        further the discussion in the commentary.

Commentary

This text provides a list of the rising time for complete signs of the zodiac. Each
entry is similar to the summary statements about the rising times for a sign of the
zodiac in the microzodiac texts. Unfortunately, entries for only the first five signs of
the zodiac are preserved in this text. In Table 4.1 I summarize the preserved entries
and compare them with entries from the microzodiac texts. Unfortunately, only the
entry for Taurus is preserved elsewhere and, as discussed above in Sect. 4.2.1 this
entry is difficult to interpret. Despite the lack of overlap between the preserved
entries on BM 36609+ and those on the microzodiac texts, as I shall discuss in
Sect. 4.5, it is certain that both types of the text present the same basic scheme. It is
therefore possible to consider all of the entries in Table 4.1 together in order to
investigate the underlying structure of the rising time scheme.

The rising times for ten signs of the zodiac can be reconstructed with confidence.
Pisces, Aries, Taurus and Gemini all appear to have rising times of 20 UŠ for the
complete sign. Leo, Virgo, Libra, Scorpio and Sagittarius have rising times of 40
UŠ. This is in agreement with Rochberg's (2004) reconstruction of the rising time
scheme as a two-zone step function from the microzodiac texts. However, the rising
time for Cancer is clearly 30 UŠ, rather than 40 UŠ as reconstructed by Rochberg.
I suggest that the 30 UŠ rising time for Cancer is because the change between 20
UŠ and 40 UŠ rising times for a zodiacal sign takes place at the midpoint, not the
beginning, of Cancer. This is in agreement with the calendar based rising time
scheme where the change between 20 UŠ and 40 UŠ rising times per month takes
place on the dates of the solstices, which are the 15th of Months IV and X. The
consequences of this assumption for understanding of the microzodiac rising time
scheme will be considered in Sect. 4.5.

Table 4.1 Rising times for complete zodiacal signs recorded on BM 36609+ and on the microzodiac texts

Sign	BM 36609+	Microzodiac texts	Culminating distance
Aries	From 10 U[Š behind] The She-Goat to The Bright Star [of] his [Bre]ast	From The Shoulder of the Panther to the Bright Star of its Breast (1/2 sign only) (BM 34716)	20 UŠ
Taurus	From The Bright Star of his Breast to 5 UŠ in front of The Knee	[From the Bright Star of] its Breast to 4 UŠ in front of The Knee (BM 34716)	$20^?$ UŠ
Gemini	From 5 UŠ not reaching The Knee to The Heel	[...]	$20^?$ UŠ
Cancer	From The Heel to 5 UŠ behind The Dusky Stars	[...]	30 UŠ
Leo	From 5 [U]Š behind The Dusky Stars [t]o 5 UŠ behind The Crook	[...]	40 UŠ
Virgo	[From 5 UŠ behind] The Crook [to 5 UŠ behind The Rear Twin]	[...]	[40 UŠ]
Libra	[...]	[Fr]om 5 UŠ behind The Rear Twin Star [t]o 5 UŠ behind The Two Stars of the Head of the Lion (BM 35456)	40 UŠ
Scorpio	[...]	[Fr]om 5 UŠ behind The 2 Stars of the Head of the Lion to 5 UŠ behind The Single Star of its Tail (A 3427)	40 UŠ
Sagittarius	[...]	From 5 UŠ behind The Single Star if its Tail to <10 UŠ behind> The Harness (A 3427)	40 UŠ
Capricorn	[...]	[...]	
Aquarius	[...]	[...]	
Pisces	[...]	From The Single Star to [10 UŠ beh]ind The Lady of Light (BM 34664)	20 UŠ

4.5 Analysis and Reconstruction of the Microzodiac Scheme

The microzodiac scheme may be split into two parts: (i) the rising time scheme which gives positions at or behind *Ziqpu* Stars which culminate at the moment when the end of a twelfth portion of a zodiacal sign rises across the eastern horizon, material which is found in both the full and the simplified form of the microzodiac scheme; and (ii) an association between the microzodiac sign and a star which expels a flare, present only in the full version of the scheme.

4.5.1 The Rising Time Scheme

Table 4.2 collects all of the attested rising time entries from the full and simplified microzodiac texts, including entries for the 12th microzodiac portion of a sign taken from the summary statements for the rising of a zodiacal sign from its beginning to its end found in the microzodiac texts themselves or in BM 36609+. These entries are given in black in the table and the source from which they have been taken is noted. To save space, I have rendered positions in sexagesimal format with UŠ as the unit.

Two conclusions can be drawn immediately from the preserved entries from the microzodiac scheme. First, there are no discrepancies between any of the preserved entries when found in different sources, with the possible exception of the twelfth and final portion of Taurus which is given as 4 UŠ in front of The Knee in one source and 5 UŠ in front of the Knee in another source (see the commentary to BM 34713 in Sect. 4.2.1); we would expect the position to be exactly at, not in front of, The Knee. This general lack of discrepancies suggests that all of the sources attest to a single scheme. This conclusion is particularly significant when we consider that we have sources from both Babylon and Uruk. Secondly, in the four cases where an

Table 4.2 The microzodiac rising time scheme

Zodiacal sign	Portion	Culminating position	Source
Aries	1	[11;40 behind The Lady of Life]	
	2	[13;20 behind The Lady of Life]	
	3	[15 behind The Lady of Life]	
	4	[16;40 behind The Lady of Life]	
	5	[18;20 behind The Lady of Life]	
	6	At The Shoulder of the Panther	BM 34713
	7	1;40 behind The Shoulder of the Panther	BM 34713
	8	3;20 behind The Shoulder of the Panther	BM 34713
	9	5 behind The Shoulder of the Panther	BM 34713
	10	6;40 behind The Shoulder of the Panther	BM 34713
	11	8;20 behind The Shoulder of the Panther	BM 34713
	12	The Bright Star of its Breast	BM 34713; BM 36609+
Taurus	1	1;40 behind The Bright Star of its Breast	BM 34713
	2	3;20 behind The Bright Star of its Breast	BM 34713
	3	[5 behind The Bright Star of its Breast]	
	4	[6;40 behind The Bright Star of its Breast]	
	5	[8;20 behind The Bright Star of its Breast]	
	6	[10 behind The Bright Star of its Breast]	
	7	[11;40 behind The Bright Star of its Breast]	
	8	[13;20 behind The Bright Star of its Breast]	

(continued)

Table 4.2 (continued)

Zodiacal sign	Portion	Culminating position	Source
	9	[15 behind The Bright Star of its Breast]	
	10	[16;40 behind The Bright Star of its Breast]	
	11	[18;20 behind The Bright Star of its Breast]	
	12	4 in front of The Knee?	BM 34713; BM 36609+
Gemini	1	[1;40 behind The Knee]	
	2	[3;20 behind The Knee]	
	3	[5 behind The Knee]	
	4	[6;40 behind The Knee]	
	5	[8;20 behind The Knee]	
	6	[10 behind The Knee]	
	7	[11;40 behind The Knee]	
	8	[13;20 behind The Knee]	
	9	[15 behind The Knee]	
	10	[16;40 behind The Knee]	
	11	[18;20 behind The Knee]	
	12	The Heel	BM 36609+
Cancer	1	[1;40 behind The Heel]	
	2	[3;20 behind The Heel]	
	3	[5 behind The Heel]	
	4	[6;40 behind The Heel]	
	5	[8;20 behind The Heel]	
	6	[At The 4 Stars of the Horn of the Stag]	
	7	[3;20] behind The 4 [Stars of the Horn of the Stag]	BM 77242
	8	[6];40 behind The 4 Stars [of the Horn of the Stag]	BM 77242
	9	[10] behind The 4 Stars of the Horn of the [Stag]	BM 77242
	10	[1]3;20 behind The 4 Sta[rs of the Horn of the Stag]	BM 77242
	11	[1;40 behind The Dusky Stars]	
	12	5 behind The Dusky Stars	BM 36609+
Leo	1	[8;20 behind The Dusky Stars]	
	2	[11;40 behind The Dusky Stars]	
	3	[At The Bright Star of the Old Man]	
	4	[3;20 behind The Bright Star of the Old Man]	
	5	[6;40 behind The Bright Star of the Old Man]	
	6	[At Nasrapu]	

(continued)

Table 4.2 (continued)

Zodiacal sign	Portion	Culminating position	Source
	7	[3;20 behind Nasrapu]	
	8	[6;40 behind Nasrapu]	
	9	[10 behind Nasrapu]	
	10	[13;20 behind Nasrapu]	
	11	[1;40 behind The Crook]	
	12	5 behind The Crook	BM 36609+
Virgo	1	[8;20 behind The Crook]	
	2	[1;40 behind The Handle of the Crook]	
	3	[5 behind The Handle of the Crook]	
	4	[8];20 behind The Handle of the Crook	BM 35456
	5	[11];40 behind The Handle of the Crook	BM 35456
	6	[15] behind The Handle of the Crook	BM 35456
	7	18;20 behind The Handle of the Crook	BM 35456
	8	21;30 behind The Handle of the Crook	BM 35456
	9	[At] The Front Twin	BM 35456
	10	3;20 behind The Front Twin	BM 35456
	11	1;40 behind The Rear Twin	BM 35456
	12	5 behind The Rear Twin	BM 35456
Libra	1	[8];20 behind The Rear Twin	BM 35456
	2	[11];40 behind The Rear Twin	BM 35456
	3	[15 behind The Rear Twin]	
	4	[18;20 behind The Rear Twin]	
	5	[1;40 behind The Crab]	
	6	At The Rear Stars of the Crab	BM 34664
	7	[3;20 behind The Rear Stars] of the Crab	BM 34664
	8	6;40 [behind The Rear Stars of the Crab]	BM 34664
	9	10 behind The Rear Stars [of the Crab]	BM 34664
	10	13;20 behind <The Rear Stars> of the Crab	BM 34664
	11	[1;40 behind The 2] Stars of the Head of [the Lion]	BM 34664
	12	5 behind The 2 Stars of the Head of the Lion	A 3427; BM 35456
Scorpio	1	8;20 behind The 2 Stars of the Head of the Lion	A 3427
	2	1;40 behind The 4 Stars of its Breast	A 3427
	3	5 behind The 4 Stars of its Breast	A 3427
	4	8;20 behind The 4 Stars of its Breast	A 3427
	5	11;40 behind The 4 [Stars of its Breast]	A 3427
	6	15 behind The 4 Stars of it[s] Breast	A 3427
	7	18;20 behind The 4 Stars if [its] Bre[ast]	A 3427
	8	1;40 behind The 2 Stars of [its] Rump	A 3427

(continued)

Table 4.2 (continued)

Zodiacal sign	Portion	Culminating position	Source
	9	5 behind The 2 Stars of [its] Rump	A 3427
	10	8;20 behind The 2 Stars of [its] Ru[mp]	A 3427
	11	1;40 behind The Single Star of [its] Ta[il]	A 3427
	12	5 behind The Single Star of [its] Tail	A 3427
Sagittarius	1	[8;20 behind The Single Star if its Tail]	A 3427
	2	1,20 behind Eru	A 3427
	3	5 behind Eru	A 3427
	4	8;20 behind Eru	A 3427
	5	11;40 behind Eru	A 3427
	6	15 behind Eru	A 3427
	7	[18;20 behind Eru]	
	8	[21;40 behind Eru]	
	9	[At The Harness]	
	10	[3;20 behind The Harness]	
	11	[6;40 behind The Harness]	
	12	At The Yoke?	A 3427
Capricorn	1	[3;20 behind The Yoke]	
	2	[6;40 behind The Yoke]	
	3	[At The Rear Harness]	
	4	[3;20 behind The Rear Harness]	
	5	[6;40 behind The Rear Harness]	
	6	[At The Circle]	
	7	[1;40 behind The Circle]	
	8	[3;20 behind The Circle]	
	9	[5 behind The Circle]	
	10	[6;40 behind The Circle]	
	11	[8;20 behind The Circle]	
	12	[10 behind The Circle]	
Aquarius	1	[11;40 behind The Circle]	
	2	[13;20 behind The Circle]	
	3	[At behind The Star from the Doublets]	
	4	[1;40 behind The Star from the Doublets]	
	5	[3;20 behind The Star from the Doublets]	
	6	[At The Star from the Triplets]	
	7	[1;40 behind The Star from the Triplets]	
	8	[3;20 behind The Star from the Triplets]	
	9	[5 behind The Star from the Triplets]	
	10	[6;40 behind The Star from the Triplets]	
	11	[8;20] behind The Star from the Triplets	BM 46167

(continued)

Table 4.2 (continued)

Zodiacal sign	Portion	Culminating position	Source
	12	At The Single Star	BM 34664
Pisces	1	[1];40 behind The Single Star	BM 46167
	2	[3;20 behind] The Single Star	BM 46167
	3	[5 behind The Single Star]	
	4	[6;40 behind The Single Star]	
	5	[8;20 behind The Single Star]	
	6	[At The Lady of Life]	
	7	[1;40 behind The Lady of Life]	
	8	[3;20 behind The Lady of Life]	
	9	[5 behind The Lady of Life]	BM 34664
	10	6;[40] behind The Lady of Life	BM 34664
	11	8;20 behind The Lad[y of Life]	BM 34664
	12	10 behind The Lady of Life	BM 34664

entry is preserved both for the middle of a zodiacal sign (i.e. the 6th portion) and the middle of a month in the monthly schemes discussed in Chap. 3 (see Table 3.3), the culminating position is identical. The agreement between the microzodiac and the monthly scheme implies that they are both different ways of presenting the same scheme.

Returning to Table 4.2, I have restored the culminating positions for parts of the scheme where no texts are known in grey. These restored position were obtained by continuing the scheme forwards and backward from preserved entries adding either 1 UŠ 40 NINDA or 3 UŠ 20 NINDA per twelfth portion of a zodiacal sign depending upon which part of the zodiac we are in. In most cases these restorations are trivial; only when the position would reach a new *Ziqpu* Star or when the addition switched between 1 UŠ 40 NINDA and 3 UŠ 20 NINDA is there any difficultly. Except as discussed below, I have assumed that the distances between *Ziqpu* Stars are as given in the standard *Ziqpu* Star list (see Table 2.1). There are two parts of the scheme where my restoration must remain tentative. First, as already discussed, there is the problem of the entry in the summary for Taurus which gives the culminating position as either 4 UŠ or 5 UŠ in front of The Knee. In the restored scheme I have simply assumed that this is a mistake for at the Knee. Second, no evidence is preserved between the midpoint of Sagittarius and the latter part of Aquarius, except for a problematical entry for the end of Sagittarius. This absence of data is particularly unfortunate because we expect the transition point between increases in the culminating position of 3 UŠ 20 NINDA and 1 UŠ 40 NINDA, and, more significantly, because the *Ziqpu* Stars which would appear in this part of the scheme are those where the standard list of *Ziqpu* Stars and their distances from one another is very poorly attested. On the assumption that the transition between increases in the culminating position of 3 UŠ 20 NINDA and 1

UŠ 40 NINDA is at the expected midpoint of Capricorn, the culminating position should increase by 40 UŠ between the midpoint of Sagittarius and the midpoint of Capricorn. The position for the midpoint of Sagittarius (i.e. the 6th portion of Sagittarius) is 15 UŠ behind Eru. If, as I have argued above, the monthly scheme and the microzodiac scheme are two versions of the same scheme, the culminating point for the midpoint of Capricorn should be the same as the culminating point at sunrise on the 15th of Month X, which is at The Circle (see Table 3.3). Thus, we can conclude that there must be 40 UŠ between 15 UŠ behind Eru and The Circle. According to the standard *Ziqpu* Star list, there are three stars between Eru and The Circle: The Harness (sometimes called The Front Harness), The Yoke and The Rear Harness. However, the preserved sources differ on their order, sometimes switching The (Front) Harness and The Rear Harness. Furthermore, the only *Ziqpu* Star list to preserve a distances between these stars is the anomalous 26-star list, which gives distances of 25 UŠ between Eru and The (Front) Harness, 8 UŠ between The (Front) Harness and the Yoke, 9 UŠ between the Yoke and The Rear Harness, and 12 UŠ between the Rear Harness and the Circle. Following this list, the distance between 15 UŠ behind Eru and The Circle only equals 39 UŠ, not the required 40 UŠ.[7] The intervals between these stars are the only ones attested in *Ziqpu* Star lists which are not multiples of 5 UŠ. I therefore have assumed that the rising time scheme assumes that the distances between The (Front) Harness, The Yoke, The Rear Harness and The Circle are all equal to 10 UŠ. Hopefully, further fragments of the microzodiac rising time scheme will hopefully be identified in the future which will allow this assumption regarding the distances between these *Ziqpu* Stars to be confirmed or rejected. In any case, the uncertainty in restoring these entries does not affect the remainder of the scheme.

A comparison of the restored microzodiac scheme with the monthly scheme confirms that in all cases the entries for the midpoint of a zodiacal sign agrees with the entries for sunrise at the middle of a month. Thus, we can conclude that the two schemes are merely different representations of the same scheme.

4.5.2 The Month, Watch, Day and Star Scheme

The full microzodiac scheme includes statements which associate the microzodiac portion with the equivalent month, one of the three watches of the day, one of three dates (28th, 29th, and 30th) and a star which "expels a flare". The preserved entries are summarized in Table 4.3. From inspection of this table it is clear that the month is the equivalent to the minor sign of the microzodiac. For example, the 7th portion of Libra is Aries of Libra. Aries is the first zodiacal sign and so corresponds to the first month and accordingly the month given in the entry is Month I. Following the

---

[7]As discussed in Sect. 2.3, the list implies either a total of 359 UŠ or 364 UŠ, not the expected 360 UŠ.

Table 4.3 The months, watches, days and stars associated with the microzodiac scheme

Zodiacal sign	Portion	Minor sign	Month	Time period	Day	Star which expels a flare
Aries	6	Virgo	VI	morning	28	The Kidney
	7	Libra	VII	morning	28	Ninmaḫ
	8	Scorpio	VIII	morning	28	The Wolf
	9	Sagittarius	IX	morning	28	Mars
	10	Capricorn	X	morning	28	The Great One
	11	Aquarius	XI	morning	28	Numušda
	12	Pisces	XII	morning	28	The Fish
Taurus	1	Taurus	II	morning	28	The Stars
	2	Gemini	III	morning	28	The True Shepherd of Heaven
Libra	7	Aries	I	morning	[28]	[The Field]
	8	[Taurus]	II	[morning]	[28]	[The Stars]
	9	[Gemini]	III	morning	28	The True Shepherd of Anu
	10	Cancer	[IV]	[morning]	28	The Arrow
	11	[Leo]	[V]	[morning]	[28]	[The Bow]
Scorpio	1	Scorpio	[VIII]	[morning]	[28]	The Wolf
	2	[Sagittarius]	IX	morning	28	Mars
	3	[Capricorn]	X	morning	28	The Crab
	4	[Aquarius]	XI	morning	28	Numušda
	5	[Pisces]	XII	morning	28	The Fish
	6	[Aries]	I	morning	28	The Field
	7	[Taurus]	II	morning	28	The Stars
	8	[Gemini]	III	morning	28	The True Shepherd
	9	[Cancer]	IV	morning	28	The Arrow
	10	[Leo]	V	morning	28	The Bow
	11	[Virgo]	VI	morning	28	The Kidney
	12	[Libra]	VII	morning	28	Ninmaḫ
Sagittarius	1	Sagittarius	[X]	[noontime]	[29]	Scorpio
	2	[Capricorn]	X	noontime	29	The Panther
	3	[Aquarius]	XI	noontime	29	The Crab
	4	[Pisces]	XII	noontime	29	Ninmaḫ
	5	[Aries]	I	noontime	29	The Fox
	6	[Taurus]	[II]	[noontime]	[29]	[…]
Pisces	9	[Scorpio]	[VIII]	[evening]	[30]	[EN.TE.N]A.BAR.ḪUM
	10	Sagittarius	IX	evening	[30]	The King
	11	Capricorn	X	evening	30	The She-[Goat]
	12	Aquarius	XI	evening	30	The Raven

Table 4.4 Suggested assignment of the zodiacal signs to the watches of the day. Reconstructed entries are given within brackets

Watch	Zodiacal signs
Morning	Aries, Taurus, Libra, Scorpio
Noontime	[Gemini], [Cancer], Sagittarius, [Capricorn]
Afternoon	[Leo], [Virgo], [Aquarius], Pisces

month, the scheme next gives one of the three watches of the day: morning (*še-rim*), noontime (AN.NE) and afternoon (EN.USAN or KIN.SIG). The watch is governed by the major sign (Libra in the example just given). The assignments of watches to zodiacal sings cannot be fully reconstructed but since Aries, Taurus, Libra and Scorpio are all associated with the morning, Sagittarius with noontime and Pisces with evening, I suggest that the full scheme is as given in Table 4.4; the underlying rationale for assigning the watches to particular zodiacal signs is not clear to me. The scheme next associates the watches with days 28, 29 and 30, following the rule set out in BM 34713 Rev. I 12–13 at the beginning of the microzodiac scheme for Aries, namely that morning corresponds to day 28, noontime corresponds to day 29 and afternoon corresponds to day 30. Again, the rationale for these associations is obscure, but it is no doubt significant that these are the final three days of the month.

The scheme concludes with the name of a star which is said to "expel a flare". As recognized already by Schaumberger (1955), these stars are taken from the repertoire of 36 "Three Stars Each" stars. Rochberg (2004) further demonstrated that there is a direct relationship between the month, the watch/day number and the star named in each entry of the scheme, which draws directly upon the Three Stars Each tradition. The Three Stars Each texts are founded upon the assignment of one star to each of the three paths of Anu, Enlil and Ea for each month of the year. In the microzodiac scheme, the watch/day number acts as an indicator of the path (morning = Ea, noontime = Anu, afternoon = Enlil)[8] and the star is then the star from the Three Stars Each scheme for the appropriate month. The particular version of the Three Stars Each list that is used is that found on BM 34713, which also contains the copy of the microzodiac scheme for Aries edited in Sect. 4.3.1. This version of the Three Stars Each list differs from earlier lists by moving the entries in the Anu and Enlil lists forward by one month.[9] The fact that the BM 34713 version of the Three Stars Each list is used throughout the microzodiac rising time series, parts of which are preserved in several copies both from Babylon and Uruk, implies that the list on BM 34713 is not a one-off, nor is it the result of a scribal error as has sometimes been suggested, but rather represents a late form of the Three Stars Each list which became standardized.

---

[8]The rationale for this association between watch and path is not clear. For discussion of this problem, see Weidner (1967: 20–21, n. 60) and Horowitz (2014: 137–139).

[9]See Horowitz (2014: 130) for a detailed discussion of this issue.

The tablet BM 34713 is in fact extremely important for understanding the connection between the Three Stars Each tradition and the microzodiac scheme. The tablet is clearly a compilation text containing a variety of material which makes use of the Three Stars Each star lists. Unfortunately, the first and last columns of the tablet are lost. The second column on the obverse contains the Three Stars Each star lists. This list is followed by three sections each of which contain lists of omens from stars which "flare" in either the morning, noontime or afternoon watch. Following a large empty space, the tablet continues with the detailed microzodiac rising time text for Aries.[10] The stars which "flare" which are given in the omens are the same stars in the same month and watch arrangement as we find in the microzodiac texts. It seems almost certain, therefore, that the reference to stars in the microzodiac texts are to be interpreted astrologically.

4.6 Conclusion

The texts discussed in this chapter provide evidence of a complete system for the rising times of the signs of the zodiac and of their subdivisions into 12 microzodiac portions. This system is directly related to the monthly scheme giving positions which culminate at sunrise and sunset on the 15th of each month in the schematic calendar discussed in Chap. 3. As discussed in Sect. 2.4, the zodiac can be seen as a mapping of the 360-day schematic year onto the band of the zodiacal constellations and, more specifically, onto the mean or schematic motion of the Sun: over the course of the twelve months of 30 days which make up the schematic year the sun moves through the twelve 30° signs of the zodiac and therefore on the assumption that the sun's motion is uniform, it moves on 1° per day. In the schematic calendar, therefore, where the beginning of Aries is set at the beginning of the year, the position of the sun is equal to the day of the year. The fact that the calendar- and zodiac-based forms of the rising time scheme are essentially the same, and that the scheme is normed such that the transition between rising arcs of 20° and 40° per sign takes place in the middle of the signs Cancer and Capricorn, rather than the beginning as had previously been assumed, and that the middle of Cancer and Capricorn are equivalent to the dates of the summer and winter solstices (the 15th of Months IV and X in the schematic calendar), argues for the primacy of the cal-endrical form of the rising time scheme and places the scheme within the tradition of schematic astronomy.[11] I will discuss the consequences of this conclusion in Chap. 6.

---

[10]It is tempting to consider whether this blank space was intended to include a drawing that was either never inscribed, or which was drawn in ink and has now been lost. Similar blank spaces appear on BM 34719 (=LBAT 1494), a text describing the construction of several gnomon-like instruments which almost certainly was intended to have drawings.

[11]As I will discuss in Chap. 5, the so-called "GU Text" hints at the existence of the calendrical scheme before the development of the zodiac in the late fifth century BC.

Table 4.5 The complete rising time system for the midpoint of each month/sign of the zodiac

Date/zodiacal position	Culminating point	Distance to culminating point at the midpoint of the next month/zodiacal sign
I 15/Aries 15°	The Shoulder of the Panther	20
II 15/Taurus 15°	15 UŠ behind The Bright Star of its Breast	20
III 15/Gemini 15°	10 UŠ behind The Knee	20
IV 15/Cancer 15°	The 4 Stars of the Horn of the Stag	40
V 15/Leo 15°	Nasrapu	40
VI 15/Virgo 15°	15 UŠ behind The Handle of the Crook	40
VII 15/Libra 15°	5 UŠ behind The Crab = The Rear Stars of the Crab	40
VIII 15/Scorpio 15°	15 UŠ behind The 4 Stars of its Breast	40
IX 15/Sagittarius 15°	15 UŠ behind Eru	40
X 15/Capricorn 15°	The Circle	20
XI 15/Aquarius 15°	The Star from the Triplets	20
XII 15/Pisces 15°	The Lady of Life	20

Combining the evidence from both the zodiacal and the calendrical form of this scheme, it is possible to reconstruct the system in its complete form. This reconstructed system is presented in Table 4.5.

References

Brack-Bernsen L, Hunger H (2005–2006) "On the "atypical astronomical cuneiform text E": a mean-value scheme for predicting lunar latitude", Archiv für Orientforschung 51:96–107

Brown D (2005) Review of Burnett et al, Studies in the history of the exact sciences in honor of David Pingree. Wiener Zeitschrift für die Kunde des Morgenlandes 95:407–428

Horowitz W (1994) Two new ziqpu-star texts and stellar circles. J Cuneiform Stud 46:89–98

Horowitz W (2014) The three stars each: the astrolabes and related texts, Archiv für Orientforschung Beiheft 33. Berger & Söhne, Horn

Monroe MW (2016) Advice from the stars: the micro-zodiac in Seleucid Babylonia. PhD dissertation, Brown University

Rochberg F (2004) A Babylonian rising time scheme in non-tabular astronomical texts. In: Burnett C, Hogendijk JP, Plofker K, Yano M (eds) Studies in the history of the exact sciences in honour of David Pingree. Brill, Leiden, pp 56–94

Roughton NA, Steele JM, Walker CBF (2004) A late Babylonian normal and Ziqpu star text. Arch Hist Exact Sci 58:537–572

Schaumberger J (1955) Anaphora und Aufgangskalender in neuen Ziqpu-Texten. Zietschrift für Assyriologie 52:237–251

Steele JM (2014) Late Babylonian *Ziqpu*-star lists: written or remembered traditions of knowledge?. In: Bawanypeck D, Imhausen A (eds) Traditions of written knowledge in ancient Egypt and Mesopotamia, Alter Orient und Altes Testament 403 Münster: Ugarit-Verlag, pp 123–151

Steele JM, Proust C (forthcoming) Astronomical and related cuneiform texts from Uruk in the ancient orient museum of Istanbul, in preparation

Weidner E (1967) Gestirn-Darstellungen auf babylonischen Tontafeln. Österreichische Akademie der Wissenschaften, Vienna

Chapter 5
Related Texts

Abstract This chapter studies several texts which are related to the rising time schemes in order to place those schemes in a broader context. Using the fully reconstructed rising time scheme it has been possible to better understand the content of these texts. In particular, the so-called 'GU Text' is shown to include references to the rising time scheme. Since the GU Text is probably to be dated to before the development of the zodiac, the link between it and the rising time scheme confirms the priority of the calendar-based form of the scheme over the zodiac-based form, and pushes the development of the rising time scheme significantly earlier than previously assumed.

Keywords Babylon · Babylonian astronomy · Calendar · Cuneiform text · GU text · *Ziqpu* stars · Zodiac

5.1 Introduction

Several texts draw upon or present material related to the rising time scheme discussed in Chaps. 3 and 4. In this chapter I discuss four such texts. These texts are significant for understanding the place of the rising time scheme within Babylonian astronomy, a topic that will be taken up in Chap. 6.

5.2 BM 36609+ Obv

BM 36609+ is a large compilation of texts dealing with *Ziqpu* Stars and Normal Stars. The tablet is incomplete and much of the surface is badly damaged, but it appears that the obverse of the tablet contained a collection of texts dealing with *Ziqpu* Stars while the reverse contained texts concerned with Normal Stars. The tablet was edited and discussed by Roughton et al. (2004). The sections discussed

© The Author(s) 2017
J.M. Steele, *Rising Time Schemes in Babylonian Astronomy*, SpringerBriefs
in History of Science and Technology, DOI 10.1007/978-3-319-55221-7_5

below include improvements to the text based upon further collation by the present author.

I have already discussed Obv. III 14–29 which contains a list of the rising times of the complete signs of the zodiac in Sect. 4.4. Many (perhaps even all) of the other sections on the obverse are also concerned with rising times. Unfortunately, the first column and the beginning of the second column are too badly damaged to make any sense out of.

5.2.1   Obv. II 15–34

Transliteration

15   [TA 10] UŠ *a-na* 4 *šá* MÚL.LU.LIM
16   [EN 5 UŠ *ár*] *um-mu-lut* KI *šá* ALLA
17   [TA 10 UŠ] *a-na* SA$_4$ *šá* MÚL.ŠU.GI
18   [EN 5 UŠ *ár*] GÀM KI *šá* UR.A
19   [TA 5 UŠ] *ina* IGI KIŠIB GÀM
20   [EN ... UŠ *ár*] ⌐MAŠ.MAŠ⌐ KI *šá* ABSIN
21   [TA 15 UŠ *ina* IGI] ⌐ALLA⌐
22   [EN 5 UŠ *ár* 2 *šá* SAG.UR.A K]I *šá* ⌐RÍN⌐
23   [TA 5 UŠ *ina* IGI] LUGAL
24   [EN] 5$^?$ [UŠ *ár* DELE *šá* KUN-*šú* KI *šá* G]ÍR.TAB
25   TA [5 UŠ *ina* IGI *e₄-ru₆*]
26   EN 4$^?$ [...] x x [KI *šá* PA]
27   [TA ...] x [...]
28   EN [10 UŠ *ár kip-pat* KI *šá* MÁŠ]
29   TA [5] ⌐UŠ⌐ *ina* IGI [*šá maš-a-ti*]
30   EN ⌐*dele*⌐ [KI *šá* GU]
31   TA 10 UŠ *a-na* [GAŠAN.TIN]
32   ⌐EN⌐ 10 UŠ *ár* GAŠAN.TIN [KI *šá zib*]
33   TA 10 ⌐UŠ⌐ *ina* IGI⌐ [*ku-mar šá* MÚL.UD.KA.DUḪ.A]
34   [EN] ⌐SA₄⌐ [*šá* GABA-*šú* KI *šá* ḪUN]
remainder lost

Translation

15   [From 10] UŠ to The 4 Stars of the Stag
16   [to 5 UŠ behind] The Dusky Stars: Place of Cancer
17   [From 10 UŠ] to The Bright Star of the Old Man
18   [to 5 UŠ behind] The Crook: Place of Leo
19   [From 5 UŠ] in front of The Handle of the Crook

20   [to … UŠ behind] The Twins: Place of Virgo
21   [From 15 UŠ in front of] The Crab
22   [to 5 UŠ behind The 2 Stars of the Head of the Lion: Pla]ce of Libra
23   [From 5 UŠ in front of] The King
24   [to] 5$^?$ [UŠ behind The Single Star of its Tail: Place of S]corpio
25   From [5 UŠ in front of Eru]
26   to 4$^?$ [...] … [: Place of Sagittarius.]
27   [From …] … [...]
28   to [10 UŠ behind The Circle: Place of Capricorn.]
29   From [5] UŠ in front of [The Star of the Doublets]
30   to The Single Star: [Place of Aquarius.]
31   From 10 UŠ to [The Lady of Light]
32   to 10 UŠ behind The Lady of Light: [Place of Aries.]
33   From 10 UŠ in front of[The Shoulder of the Panther]
34   [to] The Bright Star [of its Breast: Place of Aries.]
remainder lost

Critical Apparatus and Philological Notes

26   The sign after EN could be 4,5 or *šá*

Commentary

As suggested in Roughton et al. (2004: 547), this section contains a list of intervals
given relative to *Ziqpu* Stars which correspond to the rising arcs of the zodiacal
signs. Following collation of the tablet and in the light of the reconstruction of the
rising arc scheme in Chap. 4 it is now possible to confirm that the intervals correspond exactly to those of the scheme. The presentation of the material is different
to other lists based upon the rising time scheme, however. Normally, positions in
rising time texts are always given as being either at or behind *Ziqpu* Stars. Here, the
position for the beginning of the sign is always given as at or in front of (written
variously as *a-na* "to" or *ina* IGI "in front of") a *Ziqpu* Star whereas the position at
the end of the sign is as usual written as either at or behind the a star. Secondly, the
rising time is labelled not by reference to the rising of the zodiacal sign but instead
by noting that the culminating interval corresponds to the "place of" a zodiacal sign.

   The list begins with the sign Cancer which corresponds to the month of the
summer solstice and whose middle marks the transition from rising times of 20 UŠ
and 40 UŠ per sign. Lists beginning with the summer solstice are common in the
genre of schematic astronomy.[1]

---

[1]See, for example, the calendrical rising time texts discussed in Chap. 3 and several of the
shadow-length scheme texts discussed by Steele (2013).

5.2.2   Obv. III 1–13

Transliteration

1   TA ⌜15⌝ MÁŠ EN 30 MÁŠ [...]
2   an-⌜na⌝-a šá ul-tu₄?
3   ṣi-i-⌜tum⌝ šá ziq-pi
4   DU?-⌜ma?⌝ BU Ú

- -

5   ⌜TA 15 ALLA⌝ EN 15 ⌜MÁŠ⌝ A.RÁ 3,20
6   [TA] ⌜15?⌝ PA EN 30? [PA] A.RÁ 5,30 3,20
7   [...] 5?
8   [...]
9   [...]
10 [...] A.RÁ x
11 TA [...] A.RÁ 2?,40 37,30
12 ⌜TA 15⌝ [MÁŠ EN] ⌜15⌝ ALLA A.RÁ 1,40
13 ⌜an-na-a⌝ [x x] x šá ziq-pi

Translation

1   From 15 Capricorn to 30 Capricorn [...]
2   that which is from
3   the rising of the *Ziqpu*
4   ...

- -

5   From 15 Cancer to 15 Capricorn, multiply by 3,20
6   [From] 15? Sagittarius to 30 [...], multiply by 5,30 3,20
7   [...] 5?
8   [...]
9   [...]
10 [...] multiply ...
11 From [...] multiply 2,40 37,30
12 From 15 [Capricorn to] 15 Cancer, multiply by 1,40
13 That ... of the *Ziqpu*

Critical Apparatus and Philological Notes

1   Roughton, Steele and Walker read the second zodiacal sign as MAŠ.MAŠ
    but collation shows that it is MÁŠ
5   Roughton, Steele and Walker read the second zodiacal sign as MAŠ.MAŠ
    but collation shows that it is MÁŠ
11 The sign sequence 2?,40 could be 2?,20 LIŠ. 37,30 could also be 30 7,30

12   The final signs were read by Roughton, Steele and Walker as 1,20 LIŠ, but colleation suggests that the LIŠ sign, which is written around the end is actually part of the number 40

Commentary

These two badly preserved sections appear directly before the list of rising times for the zodiacal signs edited and discussed in Sect. 4.4. Only the end of the first section is preserved, the beginning now lost at the end of column II. The beginning and end of the second section are preserved but the lines in the middle are very badly damaged. Lines 5 and 12 define the two halves of the zodiac from the middle of Cancer to the middle of Capricorn which have rising arcs of 20 UŠ and 40 UŠ. The figures of 3, 20 and 1, 40 in these lines are the rising arcs for each of the micro-zodiac portions of 2½°. Unfortunately, the remainder of these sections are too broken to interpret.

5.3 The GU Text

BM 78161 (88–4–19, 14), known to modern scholars as the "GU Text", has been edited and discussed by Pingree et al. (1988) and I refer the reader to Walker's edition for the text. The text contains lists of stars in "strings" (GU). The strings all contain either a *Ziqpu* Star or a position relative to a *Ziqpu* Star, usually, but not always, given at the head of the string, followed by between two and four other stars.[2] The tablet is complete but only contains entries for about two-thirds of the *Ziqpu* Stars suggesting that it is either a partial copy of the original list or, more likely, was one of two tablets in a series which contained the whole list. Walker estimates based upon the appearance of the tablet and the script that it was probably written between the seventh and the fifth centuries BC. If this dating is correct it implies that the list was compiled before the development of the zodiac.

Of relevance to the present topic of the rising time schemes are the *Ziqpu* Stars and, crucially, the distances behind *Ziqpu* Stars that have been included in the list. In Table 5.1 I summarize the *Ziqpu* material on this tablet. The text includes all of the *Ziqpu* Stars from The Front Twin[3] to The Lady of Life plus three entries which are behind an already given *Ziqpu* Star. These three entries are 5 UŠ behind The Crab, ½ *bēru* behind The 4 Stars of its Breast, and ½ *bēru* behind Eru. Comparison with the reconstructed rising time scheme shown in Table 4.5 shows that these three positions are the positions which culminate in the middle of Months VII, VIII

[2]For differing interpretations of the text, see Pingree et al. (1988) and Koch (1992). Further discussion of these interpretations may be found in Hunger and Pingree (1999: 90–100).

[3]I assume that the star The Feet and Hands of the Front Great Twin is the same as The Front Twin and that The Crook of the She Goat is the same star as The Lady of Life (on which see Roughton et al. 2004: 549).

Table 5.1 *Ziqpu* Stars and positions behind *Ziqpu* Stars which appear in the GU text

String	*Ziqpu* Star/position
B	Feet and Hands of the Front Great Twin
C	The Rear Star of the Great Twins
D	The Crab
E	5 UŠ behind The Crab
F	The 2 Stars of the Head of the Lion
G	The 4 Stars of the Breast of the Lion
H	½ *bēru* behind The 4 Stars of its Breast
I	The 2 Stars of its Rump
J	The Single Star of the Tail of the Lion
K	Eru
L	½ *bēru* behind Eru
M	The Harness
N	ŠU.PA (=The Yoke)
O	The 2nd Harness
P	The Circle
Q	The Star of the Doublets
R	The Star of the Triplets
S	The Single Star
T	The Crook of the She Goat (=The Lady of Life)

and IX. Furthermore, in the range of *Ziqpu* Stars covered on this tablet, i.e., between The Front Twin and The Lady of Life, these three cases are the only ones where the culminating position at sunrise in the middle of the month is not at a *Ziqpu* Star. It seems likely, therefore, that the scholar who composed the GU text included these positions behind *Ziqpu* Stars because they are positions which appear in the rising time scheme in the middle of months. This conclusion is particularly significant if, as it appears, the GU text is a composition dating to before the development of the zodiac because it demonstrates that the rising time scheme was originally a calendar-based scheme which related culminations to the moment of sunrise and sunset on the 15th day of the month in the schematic calendar.

5.4   BM 32276

BM 32276 (76–11–17, 2004) is a fragment from the lower left corner of a two or more column tablet (Figs. 5.1 and 5.2). A small amount of the lower edge is preserved and in the middle of the fragment only a very small amount is lost beside the left edge. The surface of the obverse is very badly abraded but enough of the text can be read to identify this part of the tablet as a copy of the microzodiac rising time text for Aries known from BM 34713 and edited above in Sect. 4.2.1. The

Fig. 5.1 BM 32276 obverse

Fig. 5.2 BM 32276 reverse

right hand columns on both the obverse and the reverse preserve only a few signs. The final column on the reverse contains a series of texts which refer to the culmination of *Ziqpu* Stars in the context of the division of the night.

Transliteration

Obv.

I

1'   [*ina muḫ-ḫi* MÚL] *ár*.MEŠ *šá* ꞌMÚL ALLAꞌ [ŠÚ-*ma* 6 DANNA...]
2'   [*u₄-mu ana* 3 ḪA].LA x ꞌKAS₅ꞌ *za-a*[*z-ma* ...]

3'   [ḪA.LA reš-tu_4 2] ⌜DANNA⌝ še-rim U[D.28 2-tu_4 ḪA. ⌜LA AN.NE⌝ [UD.29]

4'   [2 DANNA AN.NE] ⌜UD.29⌝ 3-tu_4 ḪA.LA ⌜2 KASKAL EN.USAN UD.30.KAM⌝

5'   [ki-i ina ITU.BAR KI KUR] ⌜šá ᵈUD ku-mar⌝ šá MÚL.UD.KA.⌜DUḪ.A⌝ [ana ziq]-pi DU-ma

6'   [...] x x 6-tu_4 ḪA.LA ⌜šá⌝ MÚL.ḪUN

7'   [MÚL.ABSIN šá MÚL.LU KIN] KUR ina KIN ina še-rim 28 MÚL BIR meš-ḫi

8'   [...] x x x ana ziq-pi DU-ma

9'   [šamaš KI.MIN 8-tu_4 ḪA].LA šá MÚL.LU MÚL.GÍR.TAB LU APIN KUR

10' [...] x x x x x [...]

11' [...] x x x x x [...]

12' [...] x x x x x [...]

II

1'   x [...]

2'   T[A?

3'   EN? [...]

- -

4'   x x [...]

5'   x [...]

6'   TA [...]

7'   x [...]

Rev.

I

1'   [...]

2'   [...]

3'   [...]

4'   EN⁇ [...]

5'   TA [...]

6'   (blank?) [...]

7'   x [...]

8'   x [...]

9'   TA [...]

10' [...]

II

1   [x x x x x x x x x M]ÚL.ku-mar šá MÚL.UD.KA.DUḪ.A x

2   [x x x x x x x x] ⌜6⌝ DANNA mu-ši TA ⌜ku⌝-mar

3   [šá MÚL.UD.KA.DUḪ.A EN 4 M]ÚL.MEŠ šá SI MÚL.LU.LIM EN⁇. NUN USAN MÚL.ZA⁇

4   [*a-dir* TA 4 MÚL.MEŠ *šá*] ⸢SI⸣ MÚL.LU.LIM EN 5 UŠ MÚL *rit-tu₄* GÀ
      [M]

5   [EN.NUN MURUB₄.BA x x] x *a-dir* TA 5 UŠ MÚL *rit-tu₄* GÀM

6   [EN 5 UŠ *ár* MÚL AL]LA EN.NUN UD.ZAL.LA ⸢x x⸣ x UD? *a-dir*

----

7   [ITU.APIN TA] UD.1.KAM EN UD.30.KAM TA 10 UŠ *á*[*r* S]A₄ *šá*
      GABA

8   [UD.KA].DUH.A EN UGU 3 UŠ 20 NINDA *ár* [*kin*]-*ṣa* ⸢x x⸣

----

9   [I]TU.GAN TA UD.1.KAM EN UD.30.KAM TA 3 UŠ 20 NINDA *ár* k[*in*-
      ṣ]*a* EN MÚL *a-*⸢*si-du*⸣ x

----

10   ⸢ITU.AB⸣ TA UD.1.KAM EN UD.30.KAM ⸢TA MÚL *a-si-du* EN MÚL
      um-mu⸣*-lu-tú* ⸢x x⸣

----

11   [...] ⸢x x x x⸣ [...] ⸢x x⸣

12   [...] ⸢x⸣

Translation

Obv.

I

Duplicate of BM 32276 Rev. 11–18; see the translation in Sect. 4.2.1.

II

Too broken for translation

Rev.

I

Too broken for translation

II

1   [...] The Shoulder of the Panther ...

2   [...] 6 *bēru* is the night. From The Shoulder

3   [of the Panther to The 4 S]tars of the Horn of the Stag: evening watch. ...

4   [is dark. From The 4 Stars of] The Horn of the Stag to 5 UŠ behind the
      Handle of the Cro[ok:]

5   [middle watch ...] ... is dark. From 5 UŠ behind the Handle of the Crook

6   [to 5 UŠ behind The Cr]ab: morning watch ... is dark

----

7   [Month VIII, from] the 1st day to the 30th day, from 10 UŠ be[hind The
      Bri]ght Star of the Breast

8   [of the Pan]ther to the culmination of 3 UŠ 20 NINDA behind The [Kn]ee ...

----

9   [Mon]th IX, from the 1st day to the 30th day, from 3 UŠ 20 NINDA behind The K[ne]e to The Heal ...

- -

10   Month X, from the 1st day to the 30th day, from The Heal to the Dusky Stars ...

- -

11   [...] ... [...] ...

12   [...] ...

Critical Apparatus and Philological Notes

Rev. II 3   The final sign looks like ZA, but I do not know of a star MÚL.ZA.

Rev. II 6   The broken signs could be M[ÚL x] x ŠE. UD could be read –tú at the end of a star name. Alternatively, UD could be interpreted as "day" and the final phrase translated "day is dark".

Rev. II 9   The final signs DU x encroach on the column to the right. The x is written below the DU.

Commentary

This badly damaged fragment is clearly a compilation of texts which deal with intervals which are designated in terms of the culmination of points at or behind *Ziqpu* Stars. The preserved text on the obverse is a copy of the microzodiac rising time scheme text for Aries known from BM 34713. This text started one or at most two lines above the first preserved line on the obverse, indicating that there must have been another text written before the rising time scheme.

The second column on the reverse is divided into several sections which appear to contain related texts for different months of the schematic year. The first section, which begins at the top of the column, concerns Month VII, the month of the autumnal equinox. The beginning of the section is broken but it must have contained statements of the positions at or behind *Ziqpu* Stars which culminate at sunset and sunrise, followed by a statement that the length of night is equal to 6 *bēru*. The preserved position, at The Shoulder of the Panther, must refer to sunset and is in agreement with the rising time scheme for the middle of Month VII discussed in Chap. 3. The following lines divide the night into the three night watches, each of 2 *bēru* in length and give the positions at or behind *Ziqpu* Stars which culminate at the beginning and the end of each watch. The first watch, the evening watch, is said to last from the culmination of The Shoulder of the Panther to the culmination of The 4 Stars of the Horn of the Stag. As already noted, The Shoulder of the Panther culminates at sunset in the middle of Month VII according to the rising time scheme. The 4 Stars of the Horn of the Stag is 60 UŠ behind The Shoulder of the Panther according to the *Ziqpu* Star lists, giving the correct interval of 2 *bēru* for the length of the first watch. The distances between the culminating points given for the beginning and end of the second and third watches also correctly give 2 *bēru* for the length of the watch. Following each statement of the length of the watch is a reference to something becoming being "dark" (or possibly "eclipsed"). Unfortunately, all three entries are badly damaged here and I cannot identify what it is that is dark.

The subsequent sections define intervals between culminating points that correspond to the 1st to the 30th of Months VIII, IX, and X. I have not been able to identify the phenomena corresponding to these intervals.

5.5 BM 37150

BM 37150 (80–6–17, 900) is a small fragment from Babylon preserving only a few lines on the obverse and reverse (Figs. 5.3 and 5.4). The tablet was owned by the well-known Marduk-šāpik-zēri of the Mušezib family and written by his son Iddin-Bēl. The obverse preserves only a number of unknown meaning in each line. The reverse contains a series of statements of the culminating points for the signs of the zodiac.

Fig. 5.3 BM 37150 obverse

Fig. 5.4 BM 37150 reverse

Transliteration

Obv.

1′	[...] 20[+ x ...]
2′	[...] 7,ˈ20ˈ [...]
3′	[...] 1ˈ5ˈ [...]
4′	[...] 2ˈ5ˈ [...]
5′	[...] 3,ˈ10ˈ[+ x...]
6′	[...] 2ˈ5ˈ[+ x$^?$ ...]
7′	[...] 11 [...]
8′	[...] 2ˈ4ˈ[+ x$^?$...]
9′	[...] 3,ˈ40ˈ[+ x ...]
10′	[...] ˈ4ˈ[+ x ...]

Rev.

1′	[...] x x [...]
2′	[...] ˈENˈ 10 UŠ *ár kin*-[*ṣa* ...]
3′	[... *kip*]-ˈ*pat*$^?$ˈ EN 4 *šá* MÚL.LU.LIM [...]

- -

4′	[...] ˈIMˈ ^{Id}AMAR.UD.DUB.[NUMUM A *šá* ...]
5′	[A ^{I}*mu-še*]-*zib* ŠU^{II} ^{I}MU.^{d}EN.[A A-*šú*]

Translation

Obv.

1'   [...] 20[+ x ...]
2'   [...] 7,⸢20⸣ [...]
3'   [...] 1⸢5⸣ [...]
4'   [...] 2⸢5⸣ [...]
5'   [...] 3,⸢10⸣[+ x...]
6'   [...] 2⸢5⸣[+ x? ...]
7'   [...] 11 [...]
8'   [...] 2⸢4⸣[+ x?...]
9'   [...] 3,⸢40⸣[+ x ...]
10'   [...] ⸢4⸣[+ x ...]

Rev.

1'   [...] x x [...]
2'   [...] to 10 UŠ behind The Kn[ee ...]
3'   [... The Cir]cle? to The 4 (Stars) of the Stag [...]
- -
4'   [...] Tablet of Marduk-šāpik-[zēri, son of ...]
5'   [descendent of Muše]zib. Hand of Iddin-Bē[l, his son.]

Critical Apparatus and Philological Notes

Obv. 6'   25 or 26 also possible
Obv. 8'   24, 25 or 26 are possible.
Obv. 9'   The traces after the 3 could be either the beginning of 40 or 50
Obv. 10'   Any number between 4 and 8 is possible.
Rev. 3'   Only two stacked vertical wedges of the first sign remain. It is possible that the sign could instead be a GAR sign read as the syllable -ṣa at the end of the name kin-ṣa, but this seems the less likely reading.

Commentary

The obverse contains a list of numbers after a blank space to the left. The numbers appear to be either one or two place sexagesimal numbers. The numbers increase each line up to 30 in the highest place. This suggests that the numbers either represent days and fractions of days within the schematic calendar (or perhaps *tithis* of 1/30th of a synodic month) or positions within signs of the zodiac. The latter possibility seems more likely. Unfortunately, the significance of the numbers is unknown, though it is tempting to see them as part of a star catalogue listing the positions of the Normal Stars within zodiacal signs, similar to BM 36609+ Section 8.

Only two lines of the reverse before the colophon can be read. The lines both give ranges of positions at or behind culminating stars. These second position in each entry corresponds to the position which culminates at sunrise on the 15th day of Months III and IV. It seems likely, therefore, that the first position in each line is the culminating point at sunset, and that the text is giving a list of the period from sunset to sunrise using the *Ziqpu* Stars to define the length of this period.

References

Hunger H, Pingree D (1999) Astral sciences in Mesopotamia. Brill, Leiden

Koch J (1992) Die Sternenkatalog BM 78161. Die Welt des Orients 23:39–67

Pingree D, Walker CBF (1988) A Babylonian star catalogue: BM 78161. In: Leichty E et al (eds) A scientific humanist: studies in memory of Abraham Sachs. University Museum, Philadelphia, pp 313–322

Roughton NA, Steele JM, Walker CBF (2004) A late Babylonian normal and *Ziqpu* star text. Arch Hist Exact Sci 58:537–572

Steele JM (2013) Shadow-length schemes in Babylonian astronomy. SCIAMVS 14:3–39

Chapter 6
Conclusions

Abstract This chapter considers the place of the rising time schemes within Babylonian astronomy. It argues that the schemes should be seen as part of the tradition of 'schematic astronomy' which developed out of the early astronomical text MUL.APIN. By situating the rising time schemes within the schematic astronomy tradition, it is possible to suggest that the purpose of the rising time schemes was not, as has previously been assumed, to provide a means to calculate the length of daylight, but instead to provide a description of an astronomical 'fact'. The adaptation of the rising time scheme to a zodiacal framework demonstrates that the schematic astronomy tradition is less rigid than might be supposed: instead, near concepts and ideas were assimilated within the traditional framework of schematic astronomy.

Keywords Babylonian astronomy · Description · MUL.APIN · Schematic astronomy · Theoretical astronomy · Zodiac

6.1 The Reconstructed Rising Time Scheme

The evidence presented in Chaps. 3 and 4 demonstrates the existence of a single rising time scheme that could be presented within either a calendrical or a zodiacal framework. In its calendrical form, the scheme associates points either at or behind *Ziqpu* Stars which culminate at sunset and sunrise on dates within the schematic 360 day calendar. The texts containing this scheme either give this information for the 15th day of each month or for every day of the schematic year. The scheme is normed such that it is symmetrical around the equinoxes, which take place on the 15th of Months I and VI in the schematic calendar. The date when this scheme was created is not known but if Walker's dating of the related GU Text to between the seventh and the fifth century BC is correct (Pingree and Walker 1988), it suggests that the calendrical form of the rising time scheme existed already before the development of the zodiac in the last decades of the fifth century BC.

© The Author(s) 2017
J.M. Steele, *Rising Time Schemes in Babylonian Astronomy*, SpringerBriefs
in History of Science and Technology, DOI 10.1007/978-3-319-55221-7_6

The calendar-based scheme was subsequently extended to a zodiacal framework through the one-to-one equivalence of dates in the schematic calendar with positions in the zodiac. This zodiac-based scheme was presented either by giving the range of culminating points corresponding to the rising of the beginning and the end of a zodiacal sign or by giving the culminating points corresponding to risings of a series of one-twelfth portions of a zodiacal sign (known to modern scholars as a microzodiac sign).

There seems to have been widespread knowledge of both the calendrical and the zodiacal form of the rising time scheme among scholars at both Babylon and Uruk. The calendrical form of the scheme is currently known from three different texts, one of which at least seems to have become a standard text because it is known in more than one copy. Likewise, the zodiacal form is known from several different texts. In particular, a standard composition which presented the detailed microzodiac scheme prefaced by some statements about the length of daylight and the division of the day into watches on the day of the vernal equinox, is known from several examples from Babylon and Uruk, including duplicate copies of the section for Aries from both cities.

The texts discussed in Chap. 5 demonstrate that the rising time scheme were not just texts which were copied simply for the sake of copying. Instead, these texts which draw on the rising time material and either incorporate parts of the scheme within descriptions of related astronomical phenomena or explain the mathematical rules of the scheme, show that the scheme was not only known about but also fully understood by a range of Babylonian scholars during the last few centuries BC.

6.2 The Place of the Rising Times Scheme Within Babylonian Astronomy

As first recognized by Rochberg (2004), the rising time scheme embodies the tradition of a ratio of 2:1 for the length of the longest to the shortest day which is found in MUL.APIN and other works of early Babylonian astronomy such as the Three Stars Each texts. Combined with the conclusion reached in the previous section that the scheme was almost certainly initially a calendar based scheme founded on the schematic calendar of 360 days with the solstices and equinoxes at the midpoints of Months I, IV, VII and X, this places the rising time scheme firmly with the tradition of what I term "schematic astronomy". As outlined in Sect. 2.2, schematic astronomy is founded on several key principles and parameters that are found in MUL.APIN (and other early works): the 360 day schematic calendar, the 2:1 ratio for the longest to the shortest day, the nightly increase in the duration of lunar visibility being equal to one-fifteenth of the length of night, etc.

Unlike texts concerning mathematical astronomy or goal-year astronomy, schematic astronomical texts are generally descriptive rather than procedural. The rising time texts are no exception: all of the known texts are descriptive, either simply presenting items of data in a list format or giving a series of short, third person statements that present data in prose format, accompanied by brief passages before and/or after these statements which summarize the scheme underlying this data and sometimes present related material, again all written in the third person. The descriptive nature of these texts and their connection with the broader descriptive tradition of schematic astronomy has important implications for how we understand the purpose of these texts. Schaumberger (1955) and Rochberg (2004) have interpreted the microzodiac rising time texts as either forerunners to or alternative versions of the rising time material that underlies columns C of the Systems A and B lunar theories. The purpose of this material in this interpretation, therefore, is to provide the means to calculate the length of daylight from the sun's position in the zodiac by summing the rising times for the following 180°. As Rochberg (2004: 90) astutely concluded, the resulting daylength scheme would be identical to that found in MUL.APIN. However, it should be noted that none of the microzodiac rising time texts explicitly connect the scheme to the length of daylight. The only place in these texts where the length of daylight appears is in the introductory part of the Aries section and here it is simply stated as fact that on the day of the equinox the day is equal to 6 *bēru*, or half of day plus night. The length of day given here is not connected directly to the sun's position in the zodiac, nor to the following microzodiac scheme, except in as much as the culminating point at sunrise on that date—the middle of Month I, which corresponds to the middle of Aries—is quoted along with the culminating point at sunset. Thus I would argue that the purpose of the microzodiac rising time scheme is not to provide a method to calculate the length of daylight but instead to present a description of the relationship between culminating positions and the rising of (parts of) the zodiacal signs in a way that is consistent with the schematic astronomical tradition embodied by MUL.APIN. Indeed, it may be that we should see the rising time scheme as being derived from the assumption of the 2:1 ratio for the length of daylight rather than being a tool to calculate a length of daylight according to this assumption. That is not to rule out that some Babylonian scholars may have seen the potential of the microzodiac rising time scheme to be used to determine the length of daylight and it is possible—though by no means necessary to assume—that this could have influenced the development of the method of calculating daylength in Systems A and B from the sun's position in the zodiac. But, I would argue, the purpose of the rising time scheme itself was to describe an astronomical fact, not to provide a method for calculating the length of day. The presence of astrological associations for the microzodiac portions of the zodiacal signs within the scheme also highlights that the scheme was not primarily about calculating the length of day.

The calendar-based rising time scheme provides further support for this conclusion. Again, none of the calendrical scheme texts mention the length of day or night. Instead they provide statements of the culminating points at sunrise and sunset on days in the schematic calendar. Although the length of daylight *could* be determined from the difference in culminating position at sunrise and sunset, this is not done in the texts. Furthermore, the monthly texts do not provide any information on how the culminating points on days other than the 15th could be calculated, although once the basis of the scheme is understood this could be easily done, and therefore do not provide the means to determine the length of day on days other than the 15th. Thus, if I am correct that the calendrical form of the rising time scheme existed first, it is even less likely that the rising time scheme was intended to be used to calculate the length of day.

The microzodiac texts also provide an interesting insight into the development of the schematic astronomical tradition. The underlying rules of the rising time scheme are drawn directly from MUL.APIN: the 2:1 ratio, and the schematic 360 day calendar with the placement of the solstices and equinoxes in the middle of months I, IV, VII and X. There is no new empirical data being used in creating the scheme—it is purely a mathematical development out of the basic principles of the schematic astronomical tradition. However, adapting the calendrical rising time scheme into a recently created zodiacal framework shows that the schematic astronomical tradition was more fluid than might at first be supposed. We should not be too surprised at this, especially when we recall that the zodiac itself is founded on the schematic calendar. But it is an indication that schematic astronomy was still being actively developed in the last few centuries BC and that the schematic astronomical tradition interacted with, and was therefore perhaps less separate in the minds of the Babylonian scholars from, the traditions of mathematical, goal-year and observational astronomy. It is clear that the Babylonian scholars knew that what they were doing when they were expanding upon ideas from MUL.APIN was different to what they were doing when they were developing, for example, lunar System A from empirical data and complex mathematical analysis. But these activities were not viewed as being in competition, nor was one seemingly given higher status over the other. Instead, these different facets of Babylonian astronomy were probably viewed as being complementary.

The realization that the purpose of the rising time schemes, and indeed of schematic astronomy more generally, is not to provide a means of astronomical calculation but rather to provide a description of astronomical facts—perhaps one could even say a description of how the universe is structured and behaves—is also highly significant for how we understand Babylonian astronomy more generally as a scientific and intellectual endeavor. Babylonian astronomy has been criticized by some (largely ill-informed) historians of science as not being theoretical and not interested in explanation—some going so far as to characterize it as not quite being science.[1] The reality, however, is that the theoretical and explanatory aspects of Babylonian astronomy lie to a large extent hidden under the surface of the texts. I would contend that in the schematic astronomical tradition we see clear evidence

---

[1]For a detailed analysis of the reception of Babylonian astronomy among the broad history of science community, see Rochberg (2002).

of theory and explanation. These theories and explanations were only one of a range of different theories and explanations—not always, at least in our framework but perhaps from a Babylonian perspective, consistent with one another—that underlie different aspects of Babylonian astronomy.

References

Pingree D, Walker CBF (1988) A Babylonian star catalogue: BM 78161. In: Leichty E et al (eds) A scientific humanist: studies in memory of Abraham Sachs. University Museum, Philadelphia, pp 313–322

Rochberg F (2002) A consideration of Babylonian astronomy within the historiography of science. Stud Hist Philos Sci 33:661–684

Rochberg F (2004) A Babylonian rising time scheme in non-tabular astronomical texts. In: Burnett C, Hogendijk JP, Plofker K, Yano M (eds) Studies in the history of the exact sciences in honour of David Pingree. Brill, Leiden, pp 56–94

Schaumberger J (1955) Anaphora und Aufgangskalender in neuen Ziqpu-Texten. Zietschrift für Assyriologie 52:237–251

rinted in the United States
by Bookmasters

Printed in the United States
By Bookmasters